(Re-)konstruktionen – Internationale und Globale Studien

Reihe herausgegeben von

Wolfgang Gieler, Angewandte Sozialwissenschaften, FH Dortmund, Dortmund, Deutschland

Meik Nowak, Gustav – Stresemann – Institut e.V., Bonn, Deutschland

In der Schriftenreihe werden sowohl theoretische als auch anwendungsorientierte politische, soziale, kulturelle, geschichtliche und wirtschaftliche Themen in, mit und aus Ländern des Globalen Nordens und Südens veröffentlicht. Im Fokus der Analysen liegen der internationale Vergleich und die globalen Interdependenzen. Die Reihe ist offen sowohl für Monographien und Sammelbände als auch für herausragende Qualifikationsarbeiten (Dissertationen, Habilitationen) aus den Geistes- und Sozialwissenschaften. Sie dient als Forum zur Publikation ausgewählter Studien unter anderen zu Formen der kulturellen Globalisierung, der Migrationsbewegungen, dem Umgang mit „Anderen", der Geopolitik und des globalen Klima- und Umweltwandels.

Die Praxis der Beziehungen zwischen dem Globalen Norden und Globalen Süden, die sich in der Bipolarität zwischen westlicher „entwickelten Geberstaaten" und „(wirtschaftlich) unterentwickelten Nehmerstaaten" als Rezipienten abspielt, ist das Ergebnis kulturellen (und wissenschaftstheoretischen) Vormachtdenkens des Westens. Im Rahmen der Schriftenreihe soll die Eigenständigkeit und Gleichberechtigung der Staatenwelt des Globalen Südens wahrgenommen werden. Daher ist beabsichtigt durch einen interdisziplinär und transkulturell orientierten Ansatz zu einer erweiterten Kenntnis und damit auch einer veränderten Wahrnehmung des Globalen Nordens und Globalen Südens anzuregen. Mit einer (Re-) konstruktion und Relativierung universell verstandener westlicher Wissenschaft lassen sich Konfliktfelder bestimmen, welche die Kollision der unterschiedlichen „Selbstauffassungen" aufzeigen. Somit kann ein Diskurs in Gang gesetzt werden, der zum einen etwa die „westliche" Verengung der Begriffe benennt, und zum anderen nicht-westliche Erkenntnisse als gleichrangig anerkennt, um einem ernstgemeinten Verständigungsprozess auf „Augenhöhe" zu erreichen.

Dass ein Umdenken über eine (Re-)konstruktion der Beziehungen von Globalen Norden und Globalen Süden notwendig ist, liegt auf der Hand. Wer den Versuch eines Umdenkens jedoch unternimmt, pendelt zwischen Machbarkeit und Zurückschrecken vor der Hybris. Dabei ist das eingeklammerte (Re-) zugleich ein Signal der Vorsicht und ein Herausstellen: Sich über die Konstruktion der Beziehung zu verständigen, kann nur im Rahmen einer möglichen Suche nach unterschiedlichen Sichtweisen eröffnet werden. Nicht zuletzt ist es ein Anliegen, das Bewusstsein unserer transkulturellen und umweltpolitischen Verantwortung gegenüber dem Planet Erde und dessen Bewohner*innen, gleichwohl, ob diese nun aus dem Globalen Norden oder Globalen Süden der aktuellen Weltkonstellation kommen, zu stärken. Denn Globale Interdependenzen machen nicht vor Grenzen Halt – weder vor geografischen noch vor gedanklichen.

Bedeutsam ist es daher, Wissenschaftler*innen aus dem Globalen Norden und dem Globalen Süden eine Austausch- und Diskussionsmöglichkeit zu bieten. Zudem wird selten berücksichtigt, dass auch Wissens- und Forschungspraktiken selbst in einem Kontext von politischer Gewalt-, Macht- und Herrschaftsverhältnissen sowie Rassismus stehen. Wissenschaft und Forschung sind keineswegs neutral. Eine „Entkolonialisierung der Wissenschaft" sollte demnach nicht nur als Tausch einer Gruppe von Wissenschaftler*innen für eine andere in Literaturlisten aufgefasst werden. Die (Re-)konstruktion unseres Wissens ist eine notwendige Voraussetzung, um sich aus der intellektuellen Einengung des westlichen Ethnozentrismus zu befreien. Denn „Fortschritt in eine Richtung kommt nicht ohne Aufhebung der Möglichkeit zum Fortschritt in eine andere Richtung zustande", wie es Paul Feyerabend formulierte.

Franziska Ollendorf

The Transformative Potential of Corporate Social Responsibility in the Global Cocoa-Chocolate Chain

Insights from Sustainability Certification Practices in Ghana

 Springer VS

Franziska Ollendorf
Leibniz Centre for Agricultural
Landscape Research (ZALF)
Müncheberg, Deutschland

Dissertation im Cotutelle-Verfahren an den Universitäten Gießen (Fachbereich Sozial-
und Kulturwissenschaften) sowie Toulouse (Laboratoire interdisciplinaire solidarités,
sociétés, territoires, LISST
Dissertation in cotutelle procedure at the Universities of Giessen (Department of Social
and Cultural Sciences) and Toulouse (Laboratoire interdisciplinaire solidarités, sociétés,
territoires, LISST)

ISSN 2731-0531   ISSN 2731-054X   (electronic)
(Re-)konstruktionen – Internationale und Globale Studien
ISBN 978-3-658-43667-4   ISBN 978-3-658-43668-1   (eBook)
https://doi.org/10.1007/978-3-658-43668-1

This Springer VS imprint is published by the registered company Springer Fachmedien Wiesbaden
GmbH, part of Springer Nature.
The registered company address is: Abraham-Lincoln-Str. 46, 65189 Wiesbaden, Germany

Paper in this product is recyclable.

To my family

Abstract

This study engages with the implications of an expanded Corporate Social Responsibility (CSR) of Transnational Corporations (TNCs) in their supply chains, which many practitioners and academics alike see as an important contribution to global sustainability governance. Taking the case of an UTZ cocoa sustainability certification project in Ghana, the study examines the implementation process of a transnational CSR intervention and its local outcomes. In contrast to most existing studies, this research does not look at CSR as a neutral approach of global governance but as a concept and strategy which reflects power struggles in global production fields.

The Global Cocoa-Chocolate Chain (GCCC) is facing complex sustainability challenges, most notably the extreme poverty of millions of cocoa farmers and environmental degradation in production areas. Hence, the transnational cocoa and chocolate industry is concerned with the future sustainability of its production base. Starting from this, the overarching question of the present study is on whether CSR helps TNCs to broaden their sphere of influence in their supply chains in such a way that it allows them to improve their governance and achieve control over the fragmented local production base. To answer this overriding question, a research framework has been developed which combines Global Value Chain Analysis with a neo-Gramscian perspective on global governance. This consent and control framework provides a distinct set of governmental dimensions which enable the capturing of industrial governance activities and more subtle forms of governance. The study assumes that TNCs from the GCCC use CSR to (1) intervene in the local institutional environment to improve their control over production and enable the modernization of the sector, and to (2) shape perceptions of cocoa farmers in a way that prevents them from abandoning cocoa as their first livelihood strategy.

The assessment of CSR in its current prevailing form of voluntarism is important because much hope for the improvement of social and environmental conditions in global value chains is attributed to it. But often enough, this hope lacks a sound knowledge base on the various structural effects CSR may cause in its target areas. The study seeks to contribute to the discussion of CSR as a tool for improved value chain sustainability by providing in-depth insights in transformations at the local level of the GCCC linked to it. The discussion bases on an empirical study conducted in Ghana in 2015 and 2017. One project of UTZ cocoa sustainability certification implemented by a transnational agri-food company was selected for case study. Semi-structured key informant interviews were carried out with representatives from different institutions of the cocoa industry (N=127), such as cocoa farmers – participating and not participating in the selected certification project, project implementors, representatives from the public regulatory body Ghana Cocoa Board (COCOBOD), as well as from NGOs and certification organizations, industry associations, and transnational processing and manufacturing companies. Subsequently, a qualitative content analysis was conducted and results discussed against the background of the guiding consent-and control framework.

The findings indicate that transnational CSR in the form of cocoa sustainability certification, when implemented by transnational cocoa processers, leads to an increase of TNCs governance capacities and expands their sphere of action down to the local level of cocoa farmers, to whom in the case of Ghana, they did not have a direct link to before.

Several institutional transformations resulting from the concert of many similar currently ongoing CSR projects in Ghana's cocoa sector can be observed. New forms of sector organization include a privately-run extension system, new alliances between stakeholders, new mechanisms of sector monitoring and sanctioning, extended traceability schemes, the entering of TNCs to the internal marketing segment, and new marketing relations, among others. While certified cocoa farmers enter new contractual relations with the TNCs, which in the conventional cocoa chain in Ghana do not exist, smaller local Licensed Buying Companies face pressure due to increased competitiveness linked to certification schemes. In addition, the role of COCOBOD in the sector changes and a tendential reduction of its sovereignty can be observed. While all these institutional changes potentially lead to the an improved control over local production for

TNCs, farmers' perceptions towards their future in cocoa farming seem to not to be much affected by their participation in the studied certification project. This, most likely, is due to the persistence of the severe poverty they are living in as cocoa farmers. Based on these findings, the study suggests to engage in the harmonization of multi-scale efforts to improve the sustainability of the GCCC instead of claiming for more responsibility of TNCs.

Résumé

Cette étude examine les implications de la responsabilité sociale des entreprises (RSE) des entreprises transnationales (ETN) appliquée à leurs chaînes d'approvisionnement, considérée par de nombreux praticiens et universitaires comme une contribution importante à la gouvernance mondiale de la durabilité. Prenant le cas d'un projet de certification de la durabilité du cacao UTZ au Ghana, l'étude examine le processus de mise en œuvre d'une intervention de RSE transnationale et ses résultats localement. Contrairement à la plupart des études existantes, cette recherche ne considère pas la RSE comme une approche neutre de la gouvernance mondiale, mais comme un concept et une stratégie qui reflètent les luttes de pouvoir dans ce domaine.

La chaîne mondiale du cacao et du chocolat (CMCC) est confrontée à des défis complexes en matière de durabilité, notamment l'extrême pauvreté de millions de planteurs de cacao et la dégradation de l'environnement dans les zones de production. C'est pourquoi l'industrie transnationale du cacao et du chocolat se préoccupe de la durabilité future de sa base de production à savoir les fèves de cacao. En partant de ce constat, la question primordiale de la présente étude est de comprendre si la RSE aide les ETN à élargir leur sphère d'influence dans leurs chaînes d'approvisionnement de manière à leur permettre d'améliorer leur gouvernance et de contrôler une base de production locale et fragmentée. Pour répondre à cette question primordiale, un cadre d'analyse a été élaboré qui combine l'analyse de la chaîne de valeur mondiale avec une perspective néo-gramscienne de la gouvernance mondiale. Ce cadre de consentement et de contrôle fournit un ensemble distinct de dimensions gouvernementales qui permet de saisir les activités de gouvernance industrielle et les formes plus subtiles de gouvernance. L'étude suppose que les ETN de la CMCC utilisent la RSE pour (1) modifier l'environnement institutionnel local d'une manière qui améliore leur contrôle sur la production et leur permette la modernisation du secteur, et (2) façonner les perceptions des planteurs de cacao d'une manière qui empêche leur réorientation vers d'autres stratégies de subsistance.

L'évaluation de la RSE dans sa forme de volontarisme la plus répandue est importante car on lui attribue beaucoup d'espoir pour l'amélioration des conditions sociales et environnementales dans les chaînes de valeur mondiales. Mais bien souvent, cet espoir manque d'une base de connaissances solide sur les divers effets structurels que la RSE peut provoquer dans ses domaines cibles. L'étude vise à contribuer au débat critique sur la RSE en tant qu'outil d'amélioration de la durabilité des chaînes de valeur en fournissant des données approfondies sur les transformations des CMCC opérées au niveau local. La discussion se fonde sur une étude empirique menée au Ghana entre 2015 et 2017 qui s'appuie sur l'analyse d'un projet de certification de durabilité du cacao UTZ mis en œuvre par une entreprise agroalimentaire transnationale. Des entretiens semi-structurés (N=127) ont été menés avec des représentants de différentes institutions de l'industrie du cacao, tels que des planteurs de cacao - participant et ne participant pas au projet de certification sélectionné, des responsables de la mise en œuvre du projet, des représentants du Ghana Cocoa Board (COCOBOD), des ONG et des organismes de certification, ainsi que des associations industrielles et des entreprises de transformation transnationaux. Une analyse qualitative du contenu a ensuite été réalisée et les résultats ont été discutés dans le contexte du cadre d'analyse du "consentement et contrôle".

Les résultats indiquent que la RSE transnationale sous forme de certification de la durabilité du cacao, lorsqu'elle est mise en œuvre par les broyeurs transnationaux de cacao, contribue à accroître les capacités de gouvernance des ETN et à atteindre le niveau local des planteurs de cacao, avec lesquels, dans le cas du Ghana, ils n'avaient pas de lien direct auparavant. On peut observer plusieurs transformations institutionnelles résultant de nombreux projets de RSE similaires actuellement en cours dans le secteur du cacao au Ghana. Parmi les nouvelles formes d'organisation de la filière, on peut citer entre autres, un système de vulgarisation agricole géré par le secteur privé, de nouvelles alliances entre les parties prenantes, de nouveaux mécanismes de contrôle et de sanction de la filière, des systèmes de traçabilité étendus, l'entrée des ETN dans le segment de la commercialisation interne et de nouvelles relations commerciales. Alors que les planteurs de cacao nouent des relations contractuelles avec les ETN, qui n'existent pas dans la chaîne conventionnelle du cacao au Ghana, les petites entreprises locales d'achat sous licence subissent des pressions en raison de la compétitivité accrue liée aux systèmes de certification. En outre, le rôle du

COCOBOD dans le secteur se modifie et on observe une tendance à la réduction de sa souveraineté. Alors que l'ensemble de ces changements institutionnels favorisent généralement l'exercice d'un contrôle accru par les ETN sur la production locale, la perception des agriculteurs quant à leur avenir dans la culture de cacao semble peu affectée par leur participation au projet de certification. Cela s'explique par la persistance de la grande pauvreté dans laquelle ces derniers vivent en tant que planteurs de cacao. Sur la base de ces résultats, l'étude suggère de s'engager dans l'harmonisation des efforts multi-échelles pour améliorer la durabilité de la CMCC au lieu de revendiquer une RSE volontaire des ETN.

Contents

Abbreviations and Acronyms

CAOBISCO	Chocolate, Buscuits and Confectionery of Europe
CCC	Conseil Café Cacao (Côte d'Ivoire)
CEN	European Committee for Standardization
CEPPP	Cocoa Extension Public Private Partnership
CHED	Cocoa Health and Extension Division
CMC	Cocoa Marketing Company
COCOBOD	Ghana Cocoa Board
CRIG	Cocoa Research Institute of Ghana
CSR	Corporate Social Responsibility
CSSVD	Cocoa Swollen Shoot Virus Disease
FBO	Farmer Based Organization
FFS	Farmer Field School
FLO	Fairtrade Labelling Organization International
GAP	Good Agricultural Practices
GCP	Ghana Cocoa Platform
GISCO	German Initiative on Sustainable Cocoa
GVC	Global Value Chain
GVCA	Global Value Chain Analysis
ICC	International Chamber of Commerce
ICCO	International Cocoa Organization
ICI	International Cocoa Initiative
IMF	International Monetary Fund
IMS	Internal Management System
IPE	International Political Economy
ISO	International Organization for Standardization
LBC	Licensed Buying Company

MoF	Ministry of Finance
MoFA	Ministry of Food and Agriculture
MSI	Multi-stakeholder initiative
OEIGWG	Open Ended Inter-governmental Working Group
PC	Purchasing Clerk
PCU	Project Control Unit
PPE	Personal Protection Equipment
PPRC	Producer Price Review Committee
QCC	Quality Control Company
RSC	Rural Service Centre
SAP	Structural Adjustment Programs
SPD	Seed Production Division
STCP	Sustainable Trees Crop Program
TNC	Transnational Corporation
WCF	World Cocoa Foundation
WST	World System Theory

List of Figures

List of Tables

List of Maps

Introduction

"Certification is basically a method in order to achieve sustainability. So the main purpose of certification was to encourage farmers to stay on their farm and to keep on producing the cocoa, because of looking at the challenges or the fear that farmers might migrate into other businesses or migrating out of those cocoa growing areas to cities to get a better living." (Interview with a staff member of the studied TNC, Kumasi, Ghana, 2015)

"Then he said that we should follow the instructions they give us about the cocoa otherwise there will come a time that when your cocoa has not been certified, no one will buy it." (Interview with a purchasing clerk participating in an UTZ cocoa certification project organized by the studied TNC, 2015)

"The giving of the overall and the other equipment was what made them come and join. Especially the bonus. It really pulled people into the group." (Interview with a cocoa farmer participating in an UTZ cocoa certification project organized by the studied TNC, 2015)

In times of a plurality of global crises such as environmental degradation and continuous extreme poverty, the search for strong players with the capacity and resources to intervene is wide-spread among academics and practitioners. In this context, transnational corporations (TNCs) are often presented as such strong players, since their value chains[1] reach around the globe and down to local levels

---

[1] There is a wider range of similar terms in the literature which seek to capture what can be summarized as "a set of legally independent companies linked together by recurring market operations of supply, production and distribution", Acquier et al. 2011. For instance, the terms of global value chains, production networks, commodity or supply chains represent different theoretical concepts within the field of chain research. Chapter 3 provides an entry to these different currents. The term global value chain is mainly used in this study. Global value chain research puts an emphasis on a constellation of globally dispersed economic activities

F. Ollendorf, *The Transformative Potential of Corporate Social Responsibility in the Global Cocoa-Chocolate Chain*, (Re-)konstruktionen – Internationale und Globale Studien, https://doi.org/10.1007/978-3-658-43668-1_1

in many countries, but importantly in those with weaker governmental structures, too (Matten and Crane 2005, p. 172). It is mainly the perceived absence of governmental capacity of nation states and the international community to appropriately ensure the realization of human rights and the sustainable development goals that caused the call for and implementation of more voluntary responsibility of TNCs from the 1990 s onwards (Utting and Marques 2013). What dynamics may stem from voluntary corporate social responsibility (CSR) in a given global value chain is the main topic of this study. The Global Cocoa-Chocolate Chain (GCCC) is an action field which has been strongly targeted by transnational CSR over the past decade. It was therefore selected for the detailed study and an extensive qualitative field work was conducted in Ghana's cocoa sector.

Looking at the scope and rationale for CSR of TNCs, a normative shift can be observed: Initially, TNCs were mainly called to refrain from damaging practices and human rights abuses in their supply chains, hence their responsibility to assume the negative duty to respect. But over the past two decades, the expectations towards them have broadened. By late, also positive duties, i.e. the responsibility to do good, are increasingly ascribed to TNCs by both academics (e.g. Voegtlin and Scherer 2017; Scherer et al. 2016; Crane et al. 2009) and civil society groups (see the case of the Make Chocolate Fair movement below). This is reflected in the broader call for a global governance constellation, where TNCs are regarded to form one central actor group within a multi-stakeholder constellation and are encouraged to proactively contribute to the provision of rules as well as public goods and services (Abbott and Snidal 2010; Eberlein et al. 2014; Börzel and Risse 2010; Mende 2020; Ougaard and Leander 2012). However, this practice has already mainstreamed for at least two decades. TNCs undertake governance and rule setting efforts in their supply chains through a multitude of regulatory instruments and their participation in multi-stakeholder initiatives (MSIs) (Abbott 2012; Clapp and Fuchs 2009; Hatanaka and Konefal 2013; Schwindenhammer 2017). Such practices have been already shaping the life of workers and people around the globe and still continue to do so (Kreide 2019).

It is particularly in the field of sustainability governance, that TNCs became ever more active in their governmental efforts; and CSR as an umbrella function has consolidated this dynamic. In the frame of CSR, TNCs have become major partners in many important development partnerships and are thereby directly involved in negotiating policies or the implementation of programs (cf. Beisheim

---

with the main feature of being centrally governed by lead firms, Gereffi et al. 2005; Ponte and Sturgeon 2014.

and Ellersiek 2017). For instance, the Agenda 2030 for Sustainable Development confirms the path already embarked upon at the Rio Conference in 1992 of strengthening partnerships with the private sector, which with the SDG 17 are even anchored in the set of sustainable development goals (United Nations 2015). Corresponding forums for public-private exchange, such as the Partnership Exchange forum, part of the High-level Political Forum on Sustainable Development, led by the United Nations Department of Economic and Social Affairs, or the Partnership Forum organized annually by the United Nations Economic and Social Council, are important public-private arenas in which sustainability policy deliberations take place and agendas are prepared.

The shift regarding the understanding and practice of CSR, that is from mere negative duties to voluntary positive responsibilities, mirrors the changing political climate over the past four decades. CSR measures in the 1980 s were still reactive in nature and responded to increasing pressure from civil society groups and the threat of hard regulations from the international community (Ponte 2019, p. 13). Since then, liberalization and deregulation in many sectors triggered new forms of self-regulation and market-based approaches to global sustainability governance. These include, among others, codes of conduct, production and process standards, guidelines for management systems, and certification schemes or accountability and reporting frameworks (Almeida et al. 2012). Such approaches have since become tools for CSR implementation, too. In the mid-2000 s, a wave of sustainability initiatives occurred (Bregman 2017). It has been argued, that in practice there is no difference between CSR and sustainability management and hence, they should be regarded as synonymous to each other (Loew and Rohde 2013). The overlap becomes clear when looking at the typical fields of action that belong to the core of CSR: environmental protection as well as working conditions and human rights, both within the company and with suppliers and sub-suppliers (Loew and Rohde 2013, p. 15).

In a world economy that has evolved over the past four decades into a world of global commodity chains, the involvement of so-called lead firms, i.e. TNCs that hold a dominant position in global commodity chains, has become increasingly important. With the expansion of information and communication technologies, lead firms have outsourced many production steps (Dicken 2015) resulting in highly dispersed economic activities centrally governed by them. The development of governmental tools such as codes of conduct, production standards, and accountability and reporting frameworks, with which suppliers have to conform, was an immediate concomitant of this dynamic. When implemented, the tools overlapped with the development of similar approaches in the frame of CSR, with voluntary sustainability standards (VSS)[2] as an example. Both approaches contributed to the emergence of a highly complex and fragmented private governance structure, with CSR being an important part of it. Thereby, CSR, be it in the form of participation in policy committees or the introduction of social and environmental standards, has led to drastic expansion of the scope of action of TNCs in recent years. The opinions of how desirable this development is and in which form TNCs should exercise their social and political responsibility vary widely in theory and practice. In essence, two opposing positions can be identified: On the one hand, there is pressure for governments to retain the power to shape policy and to provide a strict regulatory and monitoring framework. On the other hand, mainly coming from business actors, there are warnings that too much bureaucracy would prevent flexibility and innovation by the private sector (Beisheim and Ellersiek 2017). While there is still disagreement about the best form or framework for CSR, TNCs have made sustainability management a mainstream element of their business strategies (Humes 2011).

In the view of the fact that cases of human rights violations or environmental pollution connected with the supply chains of lead firms continue to come to light, lead firms are driving forward their mechanisms of sustainable supply chain management in their supply chains. Examples of such problems include land evictions of small farmers in areas of palm oil plantations (cf. Pye 2017) and river pollution in the course of bauxite mining (cf. Knierzinger 2018). Sustainable supply chain management is thus often intended to address problems that arose from the dumping of national social and environmental standards. This occurs over the course of site competition for foreign direct investments between many developing and emerging countries (Anlauf and Schmalz 2019), which were actively exploited by TNCs themselves to begin with. Although the various instruments of CSR and sustainability management may have the potential to counteract the negative effects of damaging business practices by supply firms, the practice of voluntary CSR through the implementation of social and environmental standards, certification schemes and other tools leads to a significant expansion of lead firms' authority (Ponte 2019, p. 17). However, expanded authority may lead to new types of side effects, such as an increase in power concentration in supply

---

[2] The United Nations Forum on Sustainability Standards, 2013, understands VSS as "standards specifying requirements that producers, traders, manufacturers, retailers or service providers may be asked to meet, relating to a wide range of sustainability metrics, including respect for basic human rights, worker health and safety, the environmental impacts of production, community relations, land use planning and others", International Institute for Sustainable Development 2020.

chains, new constraints on supplier companies, and new private monitoring and sanction mechanisms (ibid.).

Given these dynamics, it seems to be an important analytical step to explicitly consider CSR currently practiced by TNCs as a tool for extended private governance[3], and not just as a tool for greenwashing negative reputations. It is therefore an important task to identify how these CSR mechanisms roll out in practice and what effects they bring in the targeted areas. This, however, can only be understood based on empirical insights into actual transformation processes within the various fields of action related to CSR.

This is exactly where the present study is located. Against the described background of proactive CSR, it examines CSR as a governance tool in a global value chain in which CSR and sustainability interventions have become particularly well established: the Global Cocoa Chocolate Chain (GCCC). Since the introduction of the first CSR initiatives to combat child labor on West African cocoa farms in the early 2000 s, CSR programs have been continuously expanded in this supply chain. Today, CSR initiatives are mainly implemented in the form of private sustainability certification schemes, but new and increasingly proactive approaches by lead firms in the GCCC are emerging with a focus on systematically modernizing cocoa farming. With the introduction of transnational CSR and sustainability strategies in the GCCC, new organizational structures have been created at the local level. A good understanding of precisely such processes is important for assessing the appropriateness of CSR as an approach to sustainability governance. For such an analysis, the present study focuses particularly on implementation processes at the local community level. It examines the cocoa sector in Ghana and the implementation of one UTZ cocoa sustainability certification project introduced by a transnational agricultural corporation[4]. The region of focus comprises five communities in the Western North Region, one of the most productive cocoa growing areas in the country. This study seeks to provide important insights into (1) the perceptions of participating actors such as cocoa farmers and representatives of the public agency in charge of cocoa, the Ghana Cocoa Board (COCOBOD), regarding the studied CSR intervention, and (2) institutional transformations linked to it and other similar CSR projects simultaneously taking place in the same area in order to assess general tendencies of sector transformation.

---

[3] This study exclusively focusses on CSR practices of TNCs, that is lead firms in global value chains. The results are not applicable, for instance, to CSR from small and medium entreprises.

[4] In order to assure anonymity of interviewees, the name of the company cannot be made public.

The objective of the work is thus on two levels: The first level is assessing the local effects of CSR in its current practice. The main interest here is to understand how CSR affects the perceptions of cocoa farmers in terms of cocoa production and livelihood, and what are the challenges and benefits they experience during their participation in the studied CSR intervention. With respect to institutional transformations in cocoa production in Ghana, it is analyzed what new forms of sector organization and patterns of cooperation between stakeholders emerge with the implementation of CSR. Does the large number of transnational CSR interventions currently taking place in the Ghanaian cocoa sector lead to a redistribution of tasks and responsibilities, hence power positions, among public authorities and private stakeholders? These empirical findings constitute the second level of research by providing a well-founded contribution to the overarching discussion of the desirability of voluntary CSR as a soft tool of global sustainability governance.

1.1   Problem Statement

Three quarters of global cocoa bean production come from four West African countries: Côte d'Ivoire, Ghana, Cameroon and Nigeria, with Côte d'Ivoire and Ghana being the two main cocoa-growing countries (INKOTA-netzwerk 2020a). Until recently, cocoa in West Africa is mainly produced on small-scale farms where a household cultivates mostly exclusively cocoa with a small number of subsistence crops. The production of cocoa beans is followed by two processing steps: the further processing of the beans into chocolate couverture and other intermediate products, and the manufacture of chocolate products. In the processing sector, two companies alone, the transnational agricultural corporation Cargill and the chocolate producer Barry Callebaut, are responsible for more than 70 percent of chocolate couverture production. In chocolate production, five companies (Mars Incorporated, Mondelez International, Nestlé, Ferrero and Hershey) currently have over 50 percent of the market (Fountain and Hütz-Adams 2015).

The GCCC is characterized by a variety of sustainability challenges which lie on all three levels of sustainability (i.e. economic, social, and ecological). On the economic level, the highly concentrated market creates a strong imbalance in the distribution of value along the GCCC. High inequality leads to problems at the social level such as widespread poverty among cocoa farmers and child labor (Voora et al. 2019). The poor living conditions of the farmers also make cocoa farming less attractive and lead to an outflux to other sectors, especially among the younger generation. On the ecological level, cocoa cultivation, which spread in West Africa already during the colonial period, has contributed to a drastic reduction of the rainforests in the tropical belt. The soils of existing farms are losing fertility and it is almost impossible to find new land. It is also considered likely that climate change will have a negative impact on cocoa production due to increased drought in the growing regions.

These challenges have caused lead firms from both production segments[5] to repeatedly speak of an impending cocoa crisis and the scarcity of the raw material (Wexler 2016); however the production of cocoa beans has been increasing steadily on a global scale over the past decades (International Cocoa Organization 2015; 2020d). From the early introduction in the 2000 s, CSR in the GCCC has undergone a shift from initially punctual interventions to sector-wide approaches of sustainability. Through their CSR strategies, lead firms in the chain operate in many different ways in the cocoa sectors of the producing countries. Operations include participating and implementing partnerships, conducting certification schemes, undertaking measures to improve service delivery. Ultimately, the overall goal of all these interventions is to boost the sustainability of cocoa production, with a major focus of increasing quality and quantity of cocoa production. The rationale employed for these forms of CSR interventions is that with this, cocoa farmers would be able to overcome the poverty they are living in. In this discourse, however, for many years, the problem of low farm gate prices as a main cause of poverty, has been largely neglected.

There is yet another perspective on the sustainability risk. Cocoa production is only possible in the warm and humid equatorial "cocoa belt" (Ameyaw et al. 2018) and up to 90% of the world's cocoa production is stemmed by about two million small-scale farmers who grow their cocoa on small plots of land with a typical size of 2 to 4 hectares (Hainmueller et al. 2011). Huge cocoa plantations so far only exist in Indonesia and Malaysia and don't seem to be a model which could be easily transferred to the West African or Latin American contexts (Barral and Ruf 2012). Hence, the geopolitical dimension of cocoa production becomes an increasingly pressing matter for the companies. Two main questions arise from this from a TNC perspective. First, how can a fragmented production base of small-sale farmers who are not willing anymore to content themselves with far too little incomes (0.5 US\$ in Côte d'Ivoire and 0.84 US\$ in Ghana, Fountain and Hütz-Adams 2015) and their exclusion from broader development processes

---

[5] The GCCC has been characterized as a bi-polar chain where dominant TNCs are located in the two major production segments, cocoa processing and chocolate manufacturing, Fold 2002.

be controlled? Second, how can a new generation of farmers who are increasingly mobile and more aware of alternative opportunities to cocoa farming be tied to cocoa production while keeping high corporate margins, that is not significantly changing distributional patterns in the chain?

At the other end of the GCCC, consumers from the largest consumer countries in Europe and North America have developed a certain awareness on the grievances in many global value chains, such as poor working conditions, extreme poverty and environmental degradation. Particularly for chocolate products, which are regarded to have a special emotional value attached, NGOs and activists' groups bring the problem of cocoa sustainability into debate prior to the annual holiday seasons of Christmas and Eastern. Over the past decade in Europe, a broad coalition of civil society actors has been formed. United in the "Make Chocolate Fair" movement, NGOs and other civil society groups from 17 European countries strive for the recognition of cocoa farmers' rights and urge the industry to assume their CSR. In 2015, the network was able to collect 122,826 signatures for a petition handed over to the Association of Chocolate, Biscuit and Confectionery Industries of Europe (CAOBISCO) in Brussels. The main demands were the following:

- Ensure fair payment to cocoa farmers and their workers,
- Adhere to human and labour rights along the entire cocoa supply chain and oppose exploitative child labour,
- Enhance cocoa farmers' capacities to perform sustainable and diversified farming,
- Comply with above demands by adopting an independent certification and monitoring system (INKOTA-netzwerk 2020b).

These demands are particularly interesting for the present work, as they urge TNCs to expand their responsibilities at the local level of cocoa production and to establish new control mechanisms in the GCCC.

This double pressure—the consumer awareness on ethical grievances and the need for an improved sustainability of supply—led to a rapid move of TNCs of the GCCC towards the implementation of proactive CSR strategies during the 2000 s (2021). After some years of piloting social infrastructure-oriented projects in target communities, TNCs began to implement their CSR strategies mainly through third party certification schemes. During the time of data collection, UTZ and Rainforest Alliance were the most important sustainability certification organizations in the cocoa sector and both provided sustainability standards for farm management, farming practices, social, living and environmental conditions, as well as crop specific production and post-harvest measures[6]. However, the organizations merged in 2018 and UTZ does not exist as an individual organization anymore. In 2019, the new Rainforest Alliance reached out to more than 900,000 farmers covering an area of 3 million hectares comprising the former farmers' associations of the two organizations (Rainforest Alliance 2020b). In 2016, before the merger, UTZ and Rainforest Alliance had 27% and 10,6% of the global production volume, respectively (Lernoud et al. 2018, p. 81). During the 2010 decade, the shift of CSR programs towards more sophisticated, production-oriented approaches proliferated, aiming to professionalize cocoa farmers and increase cocoa quality and quantity. Up until today, the main argument for the implementation of transnational CSR strategies in the GCCC in the form of sustainability certification programs is the improvement of smallholders' farming practices, thereby increasing the quality and quantity of their product which in turn is assumed to directly translate into a general betterment of cocoa farmers' income and thus livelihoods, which then would also lead to improved environmental protection. However, the little existing empirical evidence of the medium-acting socio-economic and environmental benefits of CSR interventions at the farm level doesn't point to a real improvement of farmers' living conditions (Ingram et al. 2018; Brako et al. 2020).

The present study seeks to contribute to an improved understanding of how these interventions affect farmers while emphasizing on the perceived benefits and challenges for them regarding their participation in the certification project. Yet, it still enlarges the scope above the farm level and seeks to shed light on the structural implications of transnational CSR at the local production level. Conducting this, Ghana is a special case because in contrast to the other West African cocoa producing countries, through its public cocoa management agency, the Ghana Cocoa Board (COCOBOD, it has a strong governmental presence in the sector. Given the high economic importance of cocoa beans production for Ghana's economy, COCOBOD is in charge of the regulation and provision of many services to cocoa farmers.). Hence, when new actors enter the local scene, which in the case of transnational CSR is mainly powerful processing companies and some manufacturers, existing sector structures and relations between stakeholders are likely affected. For example, when entering a context with a general shortage

---

[6] UTZ certified 2015 and UTZ certified 2014. The joint standard of UTZ and Rainforest Alliance is not available yet at the point of writing. Since the study takes one UTZ sustainability certification as a case, the UTZ standard before the merger is the most important for the context of the study.

of many material assets, it seems likely that TNCs easily become dominant players and replace existing market actors, given their strong resource endowment
(e.g. financial capital, technologies, and knowledge). During the implementation
of transnational CSR at the local level of cocoa production, new organizational
structures, new decision-making procedures, and new forms of input provision
arise. This shapes the sector and the roles of its stakeholders, such as COCO-
BOD and local input suppliers or buying agents. New responsibilities can be
overlapping with traditional authorities in the field leading to new power constellations in the local setting. Access to the innovation can be open for all or
restricted, having different impact on the social environment at the respective
level. In addition to the potential benefits, new inequalities between people or
regions are likely to result and need to be understood to prevent people from
becoming (more) marginalized. These are just a few scenarios which seem likely
to be linked to any kind of innovation brought to formerly largely isolated areas,
particularly when strong players such as TNCs are involved.

While it is important to develop an understanding of local changes linked to
privately-run sustainability interventions, a more abstract understanding is also
basic for the appraisal of TNCs' eligibility as local sustainability actors which
they become in the course of their CSR interventions. By describing and assessing the resulting new local structures and institutions, a better picture of changing
governance patterns of the GCCC can be gained. With the penetration into the
rural areas and cocoa producing communities, lead firms from the GCCC significantly broaden their field of action in Ghana. This in turn provides them with
many new opportunities to control the production and flow of their supply, which
may alter power and bargaining position in the chain, possibly to the detriment of
weaker stakeholders. This would be an unexpected side-effect of the apparently
existing consensus for extended private responsibilities in the GCCC between
civil society groups, the international community, and the private sector in the
cocoa sector. Moreover, it would shed a new light on the broader context of CSR
in the field of global governance and value chains.

1.2   Study Approach

1.2.1   Motivation and Rationale for the Study

This study seeks to elaborate on the transformative potential of CSR in global value chains with a particular focus on the local level of cocoa production. The analysis is backed by the consent and control framework, which has been developed in the course of the study and combines neo-Gramscian with Global Value Chain theory. The discussion bases its arguments on an empirical study of one former UTZ cocoa sustainability certification project in Ghana which has been implemented by a leading transnational cocoa processing corporation. At the center of attention are effects regarding

- institutional transformations in the GCCC, especially at the local level in Ghana, and
- cocoa farmers' perceptions towards their cocoa farming in Ghana.

The initial inspiration of this work was sourced from global justice theory. Pogge has been a pioneer in arguing for global institutional justice, which goes beyond the philanthropic justice concept that sees a primary duty to help the distant poor (Singer 2015). In contrast to such an understanding of positive justice duties, Pogge bases his arguments on the negative duty to not to harm anybody (Pogge 2005). Importantly, the participation in global institutions that appear to have detrimental outcomes to others, be it close or distanced people, will be a case of harm. Looking through these lenses, the imbalances in bargaining positions and asymmetric power relations in global institutions pose a normative challenge to global governance if they produce poverty for the disadvantaged groups (Kreide 2003; 2009). According to Pogge, global institutional arrangements could only be estimated as fair institutions if everyone who is directly or indirectly affected by them is equally represented and has the same opportunity to participate in them. Such arrangements are defined as any entity which produces rules on standards, production norms, codes of conducts or policy guidelines, and can include international organizations, multi-stakeholder forums or global value chains. Furthermore, the fairness of institutions would have to be evaluated against their effects on the realization of human rights. If human rights are violated or not respected due to the constellation of a global institutional setting, everybody who has participated in the design of the rules or who indirectly or directly benefitted from the arrangement and its outcomes, is guilty of the human right violation in question. Following Pogge, it is therefore a continuous academic task to reveal underlying relations of the causes and effects of global institutions on local outcomes, and to understand how their rules shape the human rights of the affected individuals (Pogge 2003, 2009).

Motivated by this objective to reveal the relationship between a dominant global institutional arrangement and local conditions, the concept of CSR became interesting for two reasons. First, as outlined above, CSR has become one of the main governmental soft tools in global governance, particularly for sustainability governance in global value chains. Academic work on CSR is manifold, but most of the work considers outcomes mainly from a business perspective. The question of how target areas are actually affected by them still receives little academic attention (cf. Frynas and Stephens 2015). Second, analyzing CSR as a transformative governance tool contributes to a broader understanding of the concept and its intrinsic power dynamics, which are often overlooked.

1.2.2 Theoretical Background

For the purpose of discussing CSR as a private governance tool with transformative potential in a given global value chain, theories from three main research fields have been mobilized: CSR theory, Global Value Chain Analysis, and neo-Gramscian theories.

Studying CSR theories and understanding the various facets of the concept in theory and practice is crucial for this study. A special focus is put on tracing the genesis of the theoretical concept (Carroll 1979; Frederick 1994, 1998; Lee 2008; Marens 2007) and carving out how CSR transcended from an initial moral claim on business actors to the main interest of a business case. The business case means that CSR should be simultaneously beneficial for the business entity and society, and it became an integral aspect of CSR in both theory and practice. It is even the underlying assumption of more recent approaches to CSR such as theories on political CSR (Eberlein 2019; Scherer and Palazzo 2008; Matten and Crane 2005; Mäkinen and Kasanen 2016; Voegtlin and Scherer 2017). In order to discuss CSR as an appropriate tool for sustainability governance[7], especially at the local level of global value chains, a good knowledge of the normative foundations and shifts in the concept (Shamir 2004; Gastinger and Gaggl 2012; Lee 2008; Utting and Marques 2013) is helpful.

---

[7] The study follows Ponte's term sustainability governance as the "cumulative efforts by public sector actors, corporations and business associations, NGOs and other civil society groups to address environmental challenges" Ponte 2019, p. 9.

Given the nature of the current world economy to basically operate through global value chains, a good understanding of how such chains operate is crucial for the analysis of CSR. Hence, the study benefits from a rich literature body in the field of global value chain research, comprising, among others, Bair (2009; 2011; 2005), Barrientos (2010), Fold (2001, 2002), Gereffi (1996a, 2001, 2014), Kaplinsky (et al. 2003; 2010), Nadvi (2002; 2008), and Ponte et al. (2005; 2008). In this field, some scholars particularly highlight forms of lead firms' governance in chains (Gereffi et al. 2005; Kaplinsky 2010; Ponte and Sturgeon 2014; Humphrey and Schmitz 2001). Importantly, in most cases of this research strand, governance is understood as chain internal governance exercised by lead firms. The focus is then on the structures and tools they use to steer their suppliers. Even from within this research field, an increasing number of scholars argue to broaden the understanding of governance by lead firms and to also consider their actions directed to actors outside the immediate chain (Abdulsamad et al. 2015; Bair and Palpacuer 2015). This has therefore received more attention in the present study. However, the governance debate in this theoretical context lacks important features which are needed to achieve a holistic understanding of governance aspirations of lead firms and which need to be considered when assessing CSR's transformative potential in value chains.

There are private governance functions, which are less visible and are not exclusively directed towards supply chain stakeholders' production but also towards their perceptions which guide their behaviors. Additionally, value chain external features, such as its institutional environment or the local setting in which production is embedded, are equally aspects of governmental interests. Neo-Gramscian theory, among others Bieling (2016), Candeias (2008), Cox (1983), Gill (2008; 2011), Opratko and Prausmüller (2011), Sklair (1997, 2003), Winter (2007, 2011)) is particularly suitable for studying such governance aspirations. From a neo-Gramscian standpoint of hegemony theory, CSR strategies from TNCs are interpreted as tools that help TNCs extend their spheres of influence while simultaneously seeking to incorporate fundamental critiques on the status quo of hegemonic practices (Levy and Newell 2002). The "technology of voluntarism" (Blowfield and Dolan 2008) that is introduced with CSR through numerous standards and codes, prevents the establishment of hard rules and laws which most likely would be enforced through sanctions and other regulatory measures. Therefore, from the neo-Gramscian angle, CSR has been criticized to come with the risk of losing public governmental capacities and possibilities to sanction cases of non-compliance of the voluntary self-commitments by irresponsible TNCs (Utting 2007, p. 702). Furthermore, Utting points to a tendential neglection of crucial development-intrinsic questions in policy dialogues. Structural dependencies, asymmetric power relations, and distribution mechanisms in supply chains which reproduce poverty in the long-run (Fischer 2020) are not thematized in mainstream sustainability governance while CSR gained its prominent role.

Both theory fields alone are not able to appropriately capture CSR as a tool for sustainability governance in a holistic way and with a particular emphasis on the local level of production. For example, whereas global value chain governance mainly focuses on industrial relations between value chain stakeholders, it is blind on TNCs' attempts to influence farmers in a way to remain in the sector and not to revolt against their poverty. Yet, the perspective on sustainability certification as a hegemonic power tool is able to explain TNCs' aspiration to make cocoa farming more attractive again. By contrast, a neo-Gramscian governance approach is not instructive for the explanation of production-related codes of conduct and standards, for instance. In order to capture both aspects of governance through CSR, the consent and control framework has been developed for this study. The framework builds upon the strengths of both approaches and integrates their key aspects: Input-output structure, geographical scope, institutional environment, and governance for the GVCA, and actors, scales, topic of consent creation, means of action, as well as perceptions for a neo-Gramscian interpretation of global governance. This combined set of angles allows to broaden the perspective for ongoing dynamics and to understand the multitude of TNCs' governmental needs and aspirations which target both chain internal and external actors and institutions. The consent and control framework provided an important guiding research grid during the field work as well the subsequent data analysis and discussion of findings. It is therefore part of the theoretical considerations as well as research methodology. Yet, it is only to be considered as a guiding framework since it was not the objective to testify its validity.

1.2.3   Methodological approach / proceeding of the empirical study

Following the objective of global justice theory and considering the theoretical background, this study analyses the implementation mechanisms and transformative effects of one selected CSR strategy in the GCCC at the local level in Ghana. Next to the above-mentioned objectives of the study, the superordinate interest of the case study lies in the understanding of if and how TNCs gain control over a dispersed local production base which is difficult to access thereby increasing their power in the GCCC. The improved understanding of major changes at the local level of production linked to the implementation of the sustainability intervention sheds light on the question of whether sustainability certification is an appropriate tool for TNCs to respond to the pressure coming from consumers and producers alike.

The discussion of these aspects builds upon extensive field work conducted in five cocoa producing communities in Ghana's Western North Region which were targeted by one important CSR strategy in the sector implemented by a lead firm from the processing segment. In the frame of this CSR project, the TNC established an advanced partnership with one of the biggest local Licensed Buying Companies in Ghana and an international NGO. They jointly organized an UTZ cocoa sustainability certification project from 2012 to 2016 which will be called hereafter "the studied project".

The empirical study triangulates qualitative methods and consists of two sequences. In 2015, during the first sequence of nine months, 42 semi-structured key informant interviews were conducted with stakeholders in the Ghanaian cocoa sector. The stakeholders were from various public cocoa-related institutions and NGOs, as well as national and transnational companies. An additional 48 semi-structured interviews with both participating and non-participating cocoa producers in the studied project as well as five focus group discussions were held in five target communities. During the second sequence in 2017, which took three months, a follow-up study was conducted. Then, 17 key informants from the public and private cocoa industry, mainly based in Accra, were interviewed and 10 follow-up interviews were conducted with those cocoa farmers interviewed in 2015 who were still available. The empirical insights gained were then analyzed and discussed in the light of the guiding framework. All farmers interviews and 24 interviews with the most relevant key informants were analyzed with content analysis software, while the others served for background information. Further details on the field work are provided in Chapter 6 on field work design and procedures, and a list of interviewees can be found in the Electronic Supplementary Material.

When approaching the field in 2015, the first aim was to understand the composition of the Ghanaian cocoa sector, which involved getting an overview of the stakeholders, their interests and positions in the field. By then, the picture of CSR in Ghana's cocoa sector was very fuzzy and no general overview of all existing interventions available. Therefore, it became a crucial part of the study to identify and cluster interventions according to their organizational structures and contents. The lack of structured knowledge on transnational CSR interventions in Ghana's cocoa sector was illustrated by the fact that, in 2014, the Swiss

Embassy instructed one of its trainees to draft an overview on all existing CSR and sustainability interventions in the field (Inglin 2014). Even the Ghana Cocoa Marketing Board, COCOBOD, as the main regulating institution of the Ghanaian cocoa sector, by then had only recently established the Programme Coordination Unit (PCU), which had as a main task to document and evaluate all existing private interventions at the production level. Hence, one major first working step of the study was the thorough gathering of information on CSR programs in place and a tentative classification of them, which will be topic in Subchapter 5.4.3.

Based on this overview, the studied project was selected as the case for in-depth study. The objectives here are (1) a description of implementation mechanisms and (2) understanding the resulting transformative processes. Hence, the empirical study comprises a detailed analysis of the implementation process, its mechanisms, and the resulting structure at the local level of cocoa governance. The study explores, among other aspects, which actors have been involved, who are the project participants, and how have they been selected for participation. It also considers if there are social and geographical particularities and which tools are used in the process. This mechanism-oriented information on the intervention is needed to understand how private sustainability governance works at the local level. The second aspiration of the empirical study is the estimation of local transformations, and in particular changes in the institutional environment and the perspectives and attitudes of sector stakeholders towards cocoa production and responsibilities in the sector. Since many similar CSR projects have been put in place in the same period, a broader effect of CSR can be expected. Particular emphasis is given to the formation of new institutions, ties or contractual arrangements between stakeholders, and the new responsibilities and structures that emerge. In sum, the study seeks to understand how TNCs use CSR as a governance tool at the highly fragmented production base in Ghana in order to assess CSR as a tool for sustainability governance.

1.2.4 Guiding Research Questions and Hypotheses

Given the fact that TNCs from the GCCC are dependent on millions of small-scale producers which are difficult to manage as compared to a large-farm supply base, the study is guided by theory-based assumptions that TNCs use CSR to gain ground at the local level of production. Accordingly, they seek to establish a governmental structure that enables them to secure and increase their sourcing control and thereby improve the quality and quantity of their cocoa beans supply.

Contrary to the above-mentioned arguments which point to the fact that CSR and sustainability strategies apply the same instruments and follow the same objectives, some TNCs from the GCCC distinguish between CSR and cocoa sustainability. Several interviewees mentioned that they increasingly differentiate between CSR, which embraces merely philanthropic projects, and sustainable supply chain management, which would openly follow a business case. This is somehow reflected by the tendency that most of the companies now have their sustainability departments in place, which are represented on distinguished websites while (traditional) CSR receives much less attention and almost disappears in the web presence. For big business entities such as the studied TNC, which operate in different agricultural supply chains and often base their aspirations on the Sustainable Development Goals, everything is about sustainable supply chain management now. Looking at the particular case of cocoa production and CSR in Ghana, the shift from philanthropic CSR to pro-active, ground-shaping strategic interventions, which occurred over the 2000 decade, becomes visible. The evolution of CSR interventions, and how sustainability grew out of it, will be traced in Section 5.4.

Still, cocoa sustainability projects base on the initial pillar of traditional CSR focused on the improvement of farmers' livelihoods, which was a directly resulting objective of the anti-child labor campaigns. During the past ten years, most of the companies have chosen to implement their sustainability strategies by applying either UTZ or Rainforest Alliance sustainability certification schemes and thereby follow similar approaches. Looking at the lead firms' websites, a number of main recurrent instruments and goals of their sustainability strategies can be found:

- Increase efficiency of small-scale producers to achieve better quality and quantity of beans,
- Thereby, enable cocoa farmers to attain better incomes,
- Thereby, improve their livelihoods,
- And achieve this through partnerships and support of farmer organizations.

Looking at these ambitious CSR objectives stated by the cocoa industry, main questions about their implementation arise. The overall research interest and hence overriding research question is if, and if yes how does CSR serves TNCs as a tool to expand their power position in a value chain (**research question 1**). The corresponding hypothesis 1 suggests that CSR frames TNCs' interventions in their value chains in a way that local transformations which benefit the implementing TNC can be achieved (governance through control) and justified (governance through consent).

In order to translate this overriding question to the concrete context of the GCCC, the study seeks to answer the more precise question of how and by which means the implementation of a respective CSR intervention (certification scheme) is achieved and how this new arrangement transforms the institutional setting of the Global Cocoa-Chocolate Chain at the local level in Ghana (**research question 2**).

Moreover, following one of the main theoretical assumptions, that is that the maintenance of the dominant position in the Global Cocoa-Chocolate Chain is based on the accommodation of critiques on the hegemonic status quo, the second main question that emerges is on how the CSR strategy affects the farmers' perceptions towards their livelihoods and their cocoa production, and finally, whether cocoa production becomes more attractive to them again (**research question 3**).

Having these questions in mind, the study approaches the field with two respective hypotheses, which helped to structure the working process. Regarding the GCCC, CSR is understood to be used as a tool to respond to multiple sustainability pressures in the chain and to establish a direct link with cocoa farmers in Ghana in order to increase their production capacities. The study assumes that these new interventions at the community level lead to new forms of sector governance in terms of organizational structure, responsibility, accountability and sanctioning mechanisms, and transparency which seek to support the modernization of the sector which is the major interest of TNCs in the GCCC (hypothesis 2: With their CSR interventions, TNCs transform the institutional setting at the local level of the sector in Ghana in a way that enables them to gain more *control* over the fragmented production base and implement their particular interest of professionalizing cocoa farmers and modernizing the sector). Next to this, the material improvements and new services delivered to cocoa farmers may raise the attractiveness of cocoa farming and lead to a perception of better livelihood conditions among cocoa farmers even if major patterns of value distribution along the chain remain in place. Besides, also the perceptions of other sector stakeholders, as importantly COCOBOD staff members, towards the role of TNCs at the local level will change in a way that the acceptance of a local role for TNCs increases (hypothesis 3: CSR helps TNCs to create *consent* among stakeholders for their particular interest while keeping the status quo of unfair value distribution in the chain in place).

1.3 Organization of the Study

The study is organized in two parts to reflect the divide of the research process. Part A) "CSR as a governance tool in the GCCC" sets out the theoretical underpinnings of the study and gives a detailed overview of the composition and functioning of the GCCC. This provides the base for part B) CSR and transformations in Ghana's cocoa sector. The core of part B is the unfolding of the case study in the environment of Ghana's cocoa sector.

Chapter 2 provides the background for the study by introducing the concept of CSR, its origin and important normative shifts, currently prevailing approaches and major streams of critique on it. The CSR research stream "Political CSR" gains particular attention since it has the most relevant implications for the research objective of the study. After the theoretical underpinnings, two main policy tools at the global level for CSR are presented, the UN Guiding Principles on Business and Human Rights, and the OECD Guidelines for Multinational Enterprises. Both tools function as soft regulations for TNCs at a global scale but also inform policy processes on regional and national levels. The chapter closes by giving an overview of the most influential private and multi-stakeholder CSR schemes.

In Chapter 3, the consent and control framework is unfolded, which gave guidance throughout the study of CSR in the GCCC. As highlighted earlier in this introduction, the consent and control framework brings together major features of GVCA and neo-Gramscian theory to one distinct analytical set of components. Accordingly, Chapter 3 introduces to GVCA, its evolution, and its major receptions. The understanding of governance in this research stream is emphasized due to its importance for the present study. Several critiques on the prevailing understanding of global value chain governance are presented whereby the need to further develop the understanding of governance becomes apparent, if the objective is to achieve a holistic picture of governance through CSR. The second part of Chapter 3 introduces to neo-Gramscian theory. One section is dedicated to Antonio Gramsci's life and work, followed by a section on the current main receptions of his intellectual legacy in the field of International Political Economy. Some key notions of his theory are then carved out and brought into the context of CSR in global value chains, thereby laying the basis for an understanding of consent-mechanisms of governance. Chapter 3 ends by elaborating the consent and control framework.

Chapter 4 introduces the functioning and important attributes of the GCCC, including historical and geographical features, activities in the main production segments, and currently prevailing governmental patterns. The chapter outlines the major sustainability challenges in the chain which need to be considered when appraising CSR and its various dynamics in the GCCC. The chapter ends by giving an overview on forms of CSR in the GCCC.

Chapter 5 focuses on the national context of cocoa production in Ghana, and outlines the background information of transformative dynamics at the local level in Ghana stemming from transnational CSR. The chapter gives insights in frontier dynamics in Ghana's cocoa sector and the institutional environment of cocoa production. Ghana keeps a unique form of sector governance which will be explained in detail. Finally, the most pressing sector challenges are illustrated and responses of different sector stakeholders presented.

Chapter 6 illustrates the methodology of the empirical study in four steps. First, the overall research approach is unrolled, where the qualitative research approach and mechanism-oriented explanatory strategy are explained. Secondly, the consent and control framework, developed in Chapter 3, is operationalized in variables and embedded in the guiding model for the empirical study. Thirdly, the field work strategy and the applied data collection methods, key informant interviews, focus group discussions, and non-participant observations, are described. Finally, the procedures of data handling and analysis are explained.

Chapter 7 processes the case study by first providing detailed information on the participating stakeholders and the modus operandi of the selected CSR project. It then goes on by giving information directly linked to the two research questions. The second part of Chapter 7 is dedicated to shed light on stakeholders' perceptions regarding the selected CSR intervention and the role TNCs should play at the local level of the sector. While experiences and attitudes of the three main stakeholder groups (the implementers of the studied project, the public cocoa agency COCOBOD, and the cocoa farmers) are of interest, the focus is on farmers' views, which are most directly shaped by the project. In the third part of the chapter, attention is given to institutional transformations linked to the studied project and the multitude of simultaneously conducted, similar sustainability certification projects in Ghana's cocoa sector.

In Chapter 8, the empirical findings exposed in Chapter 7 are discussed from the perspective of the consent and control framework, which answers the research questions. Furthermore, Chapter 8 places the case study and findings from the Ghanaian context back to the broader context of CSR as a sustainability tool, what is the overarching interest of the study. This then is the basis for drawing general conclusions and giving an outlook on a multiple-scale effort that would foster cocoa sustainability and enable cocoa farmers to overcome poverty.

Part I

CSR as a Governance Tool in the Global Cocoa-Chocolate Chain

The Evolution of CSR as "a Concept for All"

2

The idea of a social responsibility for business entities has been in existence almost as long as the modern form of transnational corporation. Early academic work on CSR has been traced back to US American business schools where a controversy around the consequences of the increased power of US corporations emerged during the 1930s (Blowfield and Murray 2019, p. 36). From the 1970s onwards, research on CSR has spread to other social sciences disciplines and became a prominent topic. Being closely related to ongoing economic and social globalization processes over the past decades, the concept of CSR is no longer merely something debated amongst academics. It has become a central part of business management and international policy making alike. Being widely recognized among academics, practitioners, and activists, the notion of CSR today serves as an umbrella term for different concepts and practices of corporate engagement in social, political, and environmental fields. Range, scope and moral foundation of business responsibilities and corresponding implications are still being discussed. The core of the CSR discussion, however, focuses on the extent to which market-based solutions, particularly the ones voluntary and self-regulatory in nature, are an adequate answer to social and environmental challenges caused by business activities (Blowfield and Murray 2019, p. 7).

As an academic concept, CSR has been discussed in frameworks such as corporate *responsiveness, corporate performance,* and *corporate citizenship,* as well as in broader contexts such as business and human rights, global governance, and sustainable development. There are two main strands of approaching the question of the corporate relationship with and position in society: a normative and an instrumental one. Normative debates in the fields of business ethics or political philosophy tend to discuss the sources of businesses' moral agency and what this should look like in practice. They ask whether businesses have responsibilities

F. Ollendorf, *The Transformative Potential of Corporate Social Responsibility in the Global Cocoa-Chocolate Chain*, (Re-)konstruktionen – Internationale und Globale Studien, https://doi.org/10.1007/978-3-658-43668-1_2

to consider, which are normatively less definite and often linked to notions of voluntarism or whether there are duties and obligations which are mandatory and directly linked to rights (Neuhäuser and Buddeberg 2015). Which practical implications do such moral foundations have for the relationship between corporations and society? How does this shape our expectations towards businesses' behavior regarding environmental degradation, global poverty, or other human rights abuses, for example?

Next to such normative considerations, instrumental or strategic approaches, mainly from management and organization studies, debate the question of business-societal relations. In contrast to the former, they are more concerned with the question of how corporations could perform better and which benefits would stem from this. In this academic CSR strand, the business case for CSR became prominent. These works often discuss incentives for companies to implement a CSR strategy and often seek to empirically prove their positive outcomes. Typical incentives would be improved business performance through CSR leading to an increased license to operate, improved operational efficiency, or better risk management, among others (Blowfield and Murray 2019, p. 154). Mostly, it is further argued, binding obligations would reduce companies' ability to appropriately implement their strategies. Thus, it is often claimed that CSR requirements should remain entirely voluntary and not limit TNCs activities.

Besides these two main CSR research strands, there is a wide range of CSR understandings and different concepts surrounding it. The US-American economist Votaw highlighted this particular characteristic of CSR decades ago:

> "The term [social responsibility] is a brilliant one; it means something, but not always the same thing, to everybody. To some it conveys the idea of legal responsibility or liability; to others, it means socially responsible behaviour in an ethical sense; to still others, the meaning transmitted is that of "responsible for," in a causal mode; many simply equate it with a charitable contribution; some take it to mean socially conscious; many of those who embrace it most fervently see it as a mere synonym for "legitimacy," in the context of "belonging" or being proper or valid; a few see it as a sort of fiduciary duty imposing higher standards of behaviour on businessmen than on citizens at large" (Votaw 1972, p. 25).

More than 40 years later, no single concise CSR definition has been established, (Jones 2009; Okoye 2009) and the debate on rationale, scope, and implications continues. Some scholars argue that it is time to overcome the "terminological turmoil" (Kloppers 2012, p. 106) and compromise on a core content of the CSR concept (Okoye 2009). Okoye points out that the missing definitional consensus hampers an effective institutionalization of CSR. But as Votaw's citation suggests, the convenience or the brilliance of the CSR construct might lie exactly in the absence of a concrete content determination which allows interest groups from all different positions to articulate their respective claims within the frame of CSR. CSR can therefore be understood as a locus of political negotiation, power struggles, and contestation.

In this chapter, the evolution of the theoretical approaches to CSR will be traced, whereby the most recent reception of the concept in terms of Political CSR gains particular attention. It will be shown how the instrumental view on CSR, mainly developed in business schools, became dominant and gained broader acceptance to the detriment of a normative claim of CSR. Following this, an overview of CSR's contemporary institutional embeddedness will be given, whereby the increasing claim towards TNCs to exercise more responsibility in the form of self-regulation and due diligence becomes visible. This trend, however, is backed by the mainstream CSR research which sees the future of global governance in the relational triad where national and international public representatives together with civil society actors and global business entities design, implement and monitor rules. Subsequently, the different forms of CSR in global value chains will be presented and the diversity of standards and norms as well as their importance in global governance discussed. The chapter ends by investigating the role CSR plays in development policies and their practical implementation on the ground.

2.1 CSR—the Genesis of an Ambivalent Concept

The rationale, content, and scope of CSR have changed greatly within the theoretical debate over the years. Frederick (1994) Carroll (1999), two influential CSR scholars in the field of business sciences and economics, trace an evolution of CSR theories since the first works in the 1960s. While Carroll conducts a literature analysis and portrays definitory changes, Frederick explicitly carves out a fundamental paradigm shift within CSR research which he classifies in four different CSR generations. The starting point of both influential analyses is the recognition of the ongoing controversy on the reach of businesses' social responsibility vis-à-vis the societies in which they are operating. Both authors date the beginning of an explicit academic dealing with these questions back to the 1950s. During that time, when business activities increasingly expanded around the globe, public control mechanisms weakened, mass worker and consumers protests in the US shook social stability, and the relation between American corporations and society worsened (Lee 2008, p. 57). According to Frederick and Carroll, this trend led leading business ethicists to study the implications of increasing business power and social and ecological effects of their activities.

By tracing a shift of the moral foundation of CSR, Frederick (1994) differenti- ates between a first and a second CSR generation. Accordingly, the fundamentals of the first CSR generation, developed in the 1950s and 60s, lie in a normative responsibility of large-scale enterprises for a general "social betterment" (Fred- erick 1994, p. 151). By that, responsibility is understood as an obligation of enterprises towards the society which exist independently of whether they would affect business affairs positively or negatively. In the by-then prevailing under- standing of CSR, binding statutory provisions would have to be applied to enforce them, if a business did not fulfil its societal duties (Frederick 1994, p. 151). For Bowen (1953), who is noted as one of the founding CSR researchers (Carroll 1999, p. 269), the crucial question of CSR research was about which societal responsibilities could be reasonably attributed to enterprises or business people (Bowen 1953, cited after Carroll 1999: 270). Underlying Bowen's work is the general assumption that the changes in institutional frameworks occurring in the 20th century would also call for new responsibilities of enterprises (Lee 2008, p. 57). This gave the basis for the first recognized definition in this new theory field:

> "It [the social responsibility of business people, author's note] refers to the obliga- tions of the businessmen to pursue those policies, to make those decisions, or to follow those lines of action which are desirable in terms of the objectives and values of our society" (Bowen 1953: Social Responsibilities of the Businessmen, cited after Carroll 1999: 270).

Above this general consent, the first generation of CSR research was character- ized by a lack of common understanding on what the theoretical and normative foundations of business' responsibilities are. Frederick highlights the missing common normative foundation of the concept and the neglecting of the relation- ship between business profit and social responsibility as well as a more precise operationalization and institutionalization of responsibility as the major weakness of CSR1. This is what he sees that was the reason for the falling behind of the philosophic-ethical orientation and a theoretical reframing of CSR in the 1970s.

Other CSR theorists like Marens (2007), do not attribute this internal instabil- ity of the initial CSR concept as the reason for the theoretical shift but rather the general political transitions during that period. In the 1970s, a progressive ratio- nalization of the CSR approach took place and largely replaced the normative notion by the idea of a proactive strategy of business management. The author suggests that this second generation, CSR 2, is the answer to increasing public pressures on TNCs. CSR as a concept during that time became known as a "Corporate Social Responsiveness" (Frederick 1994, p. 154 ff.; Moir 2001, p. 4 ff.). The core of CSR research became the identification of the best strategies to manage businesses' social relations and to develop businesses capacities to identify conflictual topics. The main objective was the development of management instruments and to give practical advice to enterprises on how to best respond to civil society's pressures and the threat of increased public regulations. This positivistic research strand produced a large number of empirical works which focused on the operationalization and implementation of CSR at the company level. Besides the internal management instruments for companies, the cooperation with politics gained increased attention by researchers. Such forms of cooperation would allow companies to better react to ongoing social pressures. For example, in 1971, the Research and Policy Committee of the U.S. Committee for Economic Development (CED), an influential business-led public policy organization consisting of mainly senior corporate executives, published its statement "Social Responsibility of Business Corporations". A major point of this statement is the establishment of a "Government Business Partnership for Social Progress" (Committee for Economic Development 1971, p. 50 ff.). Therewith, an important foray towards private participation in policy making and the institutionalization of public-private partnerships was achieved by corporate executives (cf. Frederick 1994: 157, Marens 2007).

The missing controversy on the normative foundation of CSR motivated Frederick to argue for a further development of CSR theories. Hence, with his call for a "Corporate Social Rectitude" (Frederick 1986) he went on to design a frame for CSR 3 which sought to establish a clear ethical conception of CSR (Moir 2001, p. 5). Only a strong normative foundation of research would allow for an appropriate and systematic evaluation of management decisions and their impacts on "...human consciousness, human community, and human continuity..." (Frederick 1986: 139). The main normative point he tried to establish with his concept of CSR 3 was the assumption "that the claims of humanizing are equal to the claims of economizing" (Frederick 1986: 139).

Yet, during the time of his endeavor, the mainstream CSR research was already focused on pragmatic management objectives. Even Carroll's "Corporate Social Performance" approach, which argues for a "total social responsibility" (Carroll 1979), including the three dimensions economic, legal, and ethical responsibility, did not effectively trigger a deeper controversy on the abstract relationship between business and society (Lee 2008: 60 ff.). The positivistic research strand remained dominant and the CSR research in the 1990s was clearly marked by strategic management approaches. Gastinger and Gaggl (2012) identify main drivers of strategic management approaches which were derived from the minimization of risks and the increase of business opportunities (Gastinger and Gaggl 2012, p. 244). Typical risk-reducing effects which were attributed to CSR in that period are the shielding of reputational damages and the preparation for future regulations. The windows of opportunities of CSR highlighted by then comprise credibility and reliance as a responsible actor in society (keyword Good Corporate Citizenship), brand loyalty through values and safety, the differentiation vis-à-vis competitors, or the opening of new market and customer segments (Gastinger and Gaggl 2012, p. 244).

With the beginning of the 21st century, this strategic understanding advanced to the "business case for CSR" (Carroll and Shabana 2010), which became the dominant approach in CSR theories. It integrates advantages of CSR from a management perspective which have been developed over the past decades into several strategic management concepts and becomes more sophisticated. The most important theoretical foundation from that perspectives are summarized by Carroll and Shabana as follows: "the reduction of cost and risk, gaining competitive advantage, developing and maintaining legitimacy and reputational capital, and achieving win-win outcomes through synergistic value creation" (Carroll and Shabana 2010, p. 85). Finally, Frederick's call for a transformation of the CSR concept, where not business but the whole cosmos with all its existing livings should be in the center of concern (Frederick 1998, p. 45), did not receive broader attention and the strategic paradigm settled on the top of the agenda.

Shamir (2010) discusses the crucial role US American business schools played in establishing the general understanding of CSR as a business tool and to neutralize its critical potential. Following Shamir, after being at the side-lines for a number of decades, the concept was rediscovered in the 1990s and been further developed as a workable management response to civil society critiques and the risk of legal enforceability. While CSR researchers developed a bulk of empirical work on best CSR practices, the initial moral claim of the concept was neglected and a general consent of enlightened self-interest anchored into the concept. A multitude of voluntary CSR techniques have been developed in business schools, such as the idea of sustainability "codes of conducts, best performance benchmarks, labelling, accreditation, certification, social auditing and reporting" (Shamir 2010: 11). The notion of voluntarism was inherent to these techniques from their invention, and binding regulation was seen as impediment to responsible practices. The result of these efforts was the emergence of a large and complex CSR industry with many new and old actors on the field (Shamir 2010).

The presented genesis of CSR within the business schools sought to uncover the paradigm shift which at least shapes three dimensions of CSR: its rationales, its frame of reference, and the forms in which it might be operationalized. First, and fundamental for the other two, the shift in rationale of businesses' social responsibility was fundamental. Where initially normative claims towards business were decisive and its contribution to the social betterment was at stake, they have been broadly replaced by a positivistic view. The impetus was no longer an ethical concern but the increasing need to respond to societal pressures and to prevent stronger regulations. Hence, the new rationale in the second generation of thought became an instrumental tool of CSR which became even more accentuated in CSR3. Second, with this, the frame of reference also shifted. While the early normative analyses were mainly interested in improving business-society relations and therefore searched for macro-level institutional solutions, the instrumentalist position predominantly engages at the organizational level of companies. They develop approaches tailored for practical company advice for managers to guide decision-making and corporate governance (cf. Lee 2008). Third, going hand in hand with the shifts in meaning and rationale, the understanding of how CSR should be established changed. The early ethical approach was based on a conception of business obligations towards society where, if necessary, government regulations would have to ensure. With the concept of CSR3, the obligation foundation of responsibility was completely replaced by the idea of a voluntary adaption of responsibility by corporate executives.

2.2   A Political Social Responsibility of Business?

The further development of the concept of CSR from a moral obligation to the business case for CSR culminated in today's prevailing understanding of CSR as a political responsibility of TNCs discussed in the frame of "Political CSR". The changing social environment of corporations and the increase of their ability to act around the globe engenders a necessity to develop a better understanding of the political aspects of corporations' activities, the pioneers of this new CSR strand argue (cf. Néron 2010, 334). In this search for an increased societal role of TNCs, some scholars see a turning back of CSR research towards a normative foundation while leaving the mere positivistic, instrumental approaches behind (Scherer and Palazzo 2011; Scherer 2018). In contrast, some others identify in the claim for a broader political role of TNCs a continuity or even an accentuation of the instrumental view of CSR (Mäkinen and Kourula 2012). In the globalized world of 21st century, there is almost no sphere of action which is not in one or another way shaped by global value chain (GVC) economies governed by TNCs. After at least three decades of CSR in practice, human rights abuses and ecological disasters linked to GVC activities are still as much reality as before. But what has changed over the past decades is the content and reach of the responsibility claims directed towards TNCs. Where initially calls focused on the omission of negative effects and the respect of human rights in their value chains, today, in both academic and civil society's debates, TNCs are asked to assume responsibility in a proactive way, to take measures of due diligence and to contribute to the provision of human rights and societal betterment. By doing so, their participation in norm-setting processes is often seen as an important step. Hence, TNCs should become a political player within a setting of power diffusion to a multitude of stakeholders.

One of the main assumptions underlying this call for or justification of new political roles of the private sector is a growing discrepancy between the international community's ability to protect its citizens from human rights violations and the increase of private sectors' abilities to act—that is to respect/disregard human rights and to add to the provision of public goods. This perceived governance gap refers to an apparent loss of governance capability and is often presented as a natural result of economic globalization. Accordingly, national governments are seen to be no longer in the position—or in some cases, particularly for developing countries, are seen to never have been in the position—to effectively oversee and regulate private activities and outcomes. Hence, the erosion of national institutions and the lacking international regulatory measures created a governance void, leading proponents of a Political CSR argue (Scherer and Palazzo 2011; Crane et al. 2009; Eberlein 2019; Moon 2002; Bartley 2007; Slager et al. 2012). According to them, in this postnational constellation of a globalized world, a multitude of transnational actors have to cooperate in order to close the governmental gap and to jointly face new challenges of global reach. Global governance enthusiasts see in this diffusion of authority to several transnational actors the realization of postnational governance without government (pioneering for this Rosenau and Czempiel 1992). Institutionalized in multi-stakeholder forums, national governments, civil society representatives, and business actors should all be working together towards the solutions of common problems (Pattberg and Widerberg 2016; Pattberg 2006).

The relational triad between government, civil society, and corporations is one of the main foundations of Scherer and Palazzo's conception of a political corporate responsibility. It is this changing interplay between governments, civil society, and corporations that the two influential authors in the development of their concept of "Political CSR" use as the main argument for the need to find a new normative base of CSR. Scherer and Palazzo identify a new demand of civil society for a social commitment of business that goes beyond their direct business activities or supply chains (Scherer and Palazzo 2008, p. 2). In the context of the transnationalization of business activities, TNCs have to become more sensitive towards civil society's demands since there are no broadly accepted legal or moral standards in the transnational sphere what makes corporations more dependent on civil society's legitimacy-ascription, they suggest (Scherer and Palazzo 2008, p. 28).The authors use civil society pressure and the apparent lack of an appropriate normative framework as the main argument for TNCs' "explicit participation in public processes of political will-formation", what they call the "politicization of the corporation" (Scherer and Palazzo 2008, p. 29). Corporations could only act responsible if they too were a legitimate part of the democratic institutions and processes (Scherer and Palazzo 2008, p. 5). They argue that there is already empirical evidence which proves corporations' assumption of enlarged responsibilities, e.g. when they take over traditional governmental responsibilities such as provision of social services in health and education, the protection of human rights, or actively engaging in self-regulation to fill regulatory gaps (Scherer and Palazzo 2008, p. 31). Thus, what the authors seek by the establishment of a Political CSR conception is a shift of an analysis from CSR as a reactive behavior of TNCs to stakeholder pressure towards an understanding of the "corporation's role in the overarching processes of (national and transnational) public will formation and their contribution to solving global environmental and social challenges" (Scherer and Palazzo 2008, p. 29).

The provision of public goods and human rights services by TNCs due to the privatization of major public responsibilities is also an important starting point for Matten and Crane's (2005) work on Political CSR. The authors developed the idea of an "extended corporate citizenship" which has been influential in the debate of Political CSR. Departing from the original responsibility of governments to respect and grant civil and political rights on the one hand, and to fulfil and protect social rights on the other hand, Matten and Crane apply two main arguments to advocate for corporate's new societal role. First, they see a shift in relations between corporations and society within the process of economic globalization. Since corporations took over many tasks and activities which were formerly reserved to be national government responsibilities, corporations' relationship with society has changed, they reason. The authors claim that this change alters corporations' citizenship situation and that the new powerful role of corporations as societal actors makes it necessary that they "have a responsibility to respect individual citizen's rights" (Matten and Crane 2005, p. 170). However, their argumentation does not end with the acknowledgement of negative duties and does not limit itself to the responsibility to respect. Secondly, they use the argument of a new societal role of TNCs to claim the need of corporations' entry to the administration of citizen's rights. The authors highlight three scenarios where, according to them, TNCs are already assuming important governmental functions, as they put it, in form of protection, facilitation, and enabling of citizen rights:

1. "where government ceases to administer citizenship" (Matten and Crane 2005, p. 172) for example after privatization or welfare reform,
2. "where government has not yet administered citizenship rights" (Matten and Crane 2005, p. 172), what, according to them, would be particularly the case in developing countries, or
3. "where the administration of citizenship rights may be beyond the reach of the nation-state government" (Matten and Crane 2005, p. 173). Here, they see a role for TNCs in actively contributing to the creation or reform processes of transnational institutions and in "administering rights where national governments cannot act effectively" (Matten and Crane 2005, p. 173).

These claims constitute a profound expansion of what is considered as CSR and it appears that therewith, the concept of Political CSR itself is highly political: It establishes a perspective on TNCs as neutral partners without any own interests (as formerly directly and indirectly acknowledged as a "enlightened self-interest"). With this understanding, CSR opens the floor for a processes of the "privatization of governance" (Melanie Coni-Zimmer 2012). Largely based on only positive pictures of empirical cases—corporations protecting human rights, providing social services, and helping to improve regulation through self-commitments, the mainstream Political CSR literature does not state anything about the vast empirical evidence from cases where corporations themselves are the source of societal and environmental damage. Furthermore, the claim of the new political role of TNCs does not take into account any driving forces of economic globalization and the distribution of benefits and burdens in this process (Mäkinen and Kourula 2012, p. 650). Interests of TNCs such as profit-maximization, the possible clash with human rights issues, and environmental protection are, for the most part, not discussed in the frame of Political CSR. But their participation in global governance institutions and processes, where rules are debated and formulated which shape the immediate environment of TNCs' and their suppliers' operations, allows TNCs to proactively advocate for their interests at these levels and can be therefore seen as part of a strategic management agenda, hence, an extended business case. In the main works on Political CSR, also little is elaborated on the risks of the evolution of transnational super corporations, increasing monopolization in many GVC, the resulting extreme power asymmetries within bargaining processes and global public-private institutions as well as on intrinsically opposing interests of stakeholders. For instance, Scherer and Palazzo's understanding of a "deliberative concept of CSR, [where] discourse quality derives from the analysis of arguments, not actors" (Scherer and Palazzo 2008, p. 30), seems unrealistic given the immense inequalities in resource endowments of the stakeholders involved in deliberation processes on multi-stakeholder platforms.

A number of critical works on Political CSR pick up these shortcomings and argue for a broader and more power sensitive understanding of CSR. Particularly, the transcendence of the private-public divide (Shamir 2010) and the turn towards a multi-stakeholder governance raises concern to some scholars. In the latter, critical voices identify a trend leading to the privatization of core public responsibilities such as the development of laws and rules (Tallontire 2007). In the global governance regime often advocated for, these instruments of coercive regulation are increasingly replaced by guidelines, principles, and other private soft law instruments (Shamir 2010). Some global governance scholars view the combination of soft and hard laws[1] as a form of legal pluralism and the diffusion of responsibilities to different major stakeholders a desired institutional arrangement which shall help to close the stated governmental gap (cf. Mende 2017). However, some others point to the risk that the diffusion of authority might lead to a sharp increase of private authority with private standards and codes of conducts being the new form of regulation. In this argumentation, a major concern is the neglecting of the search for a global institutional arrangement which would allow for holding TNCs accountable when self-commitment is not enough (Shamir 2011) and legally bind them to their human rights obligations (Kreide 2007). Above that, taking into account the resulting changing role of governments from regulators to facilitators (Shamir 2008), an economization of authority and the public sphere could in practice lead to a "market of authority" (Shamir 2008). This, however, might go hand in hand with a loss of public control over TNCs' activities when accountability and monitoring mechanisms are also privatized. But such a development would turn the basis of democratic principles upside down:

---

[1] Soft law refers to agreements, guidelines or declarations of intent that are not legally binding, as for instance resolutions of the UN General Assembly. Hard law refers to a binding legal obligation of the parties involved, which can therefore also be enforced in court, European Center for Constitutional and Human Rights e.V. 2020.

"the constitutional and law-giving power of the people to which all other powers, persons, and associations should be subject, will no longer be supreme and we face the danger that private self-regulation will become an instrument for further self-empowerment of the already powerful" (Kreide 2007, p. 18).

Above that, for some authors (e.g. Mäkinen, Kourula, Kasanen, Whelan, Shamir) it is the particular starting point of the Political CSR debate itself that is problematic. According to them, the dominant assumption that globalization would automatically come with a weakened role of the state and a therefrom derived need of increased private self-regulation and business involvement in societal governance is only one form of the possible interpretations of globalization (Mäkinen and Kourula 2012; Mäkinen and Kasanen 2015). The conception of a political role of business as the sole possible consequence of the stated global governance gap ignores the possibility of other institutional arrangements (Whelan 2012, p. 710). Therefore, the stated need of private administration of citizenship rights without exploring any other possible paths, should be understood as an advance of the neoliberal agenda since it seeks to foster the privatization of core public responsibilities (Mäkinen and Kourula 2012, p. 665). Furthermore, another conceptual obstruction stems from the broader neglecting of the motivations and interests of TNCs to adapt political CSR strategies. Whelan (2012, p. 710) highlights the tendency that in Political CSR, the nature of TNCs as highly "specialized economic organs" (Ruggy, 2008: p. 16, cited after Whelan 2012, p. 710) is completely ignored. The conception of TNCs as pro-social entities mainly goes with the exclusion of important intrinsic features such as most TNCs' need to maximize shareholders' returns, for instance. Similarly, there is little theoretical consideration of likely conflicting objectives of profit maximization and distributional equalities. Therefore, it has been argued that, from both, a descriptive and a normative perspective, TNCs' engagement into Political CSR should be understood as mainly profit-focused aiming at the promotion of their core business interests. Mäkinen and Kourula point to the fact that the distribution of

"social responsibilities and political tasks to the private firms is compatible with the political aims of extending the influence of firms and managers over their complex and unstable political and social environments" (Mäkinen, Kourula 2012, p. 666).

This, however, reveals the continuity of an instrumental approach to CSR marked by a lack of sophisticated analyses of underlying power dynamics and with insufficient attention given to societal structures within institutional processes and actor networks (cf.Banerjee 2011). It also indicates what Sabadoz (2011) calls the "functional side of CSR as a supplement for corporate profit seeking". The ambivalence of the concept of CSR that is simultaneously affirming the legitimacy of a business case for CSR and the pro-sociality of TNCs blurs the possible contradiction of both and appears to result in less consideration of the social claims of CSR. Likewise, Ungericht and Hirt (2010a, 2010b) stress the neglecting of the inherent political nature of CSR discourses in CSR research and argue for an overcoming of the apolitical interpretation of CSR. Despite the ongoing debate on TNCs' legitimacy as political actors and on the desired scope of their intervention in public responsibilities, CSR has been broadly institutionalized in main development agendas, global value chain strategies, and Human Rights instruments. In the following subchapter, a brief glimpse in contemporarily prevailing forms of CSR and their institutional embedding will be presented.

2.3 Forms and Institutional Embeddedness of CSR

CSR as a policy instrument entered the mainstream political agenda during the 1980s. Its penetration into various main public policy fields at national and international levels can be interpreted as a result of continuous advocating for CSR by both civil society and business itself. In the international development discourse, it quickly matured to a wide-spread umbrella term, much in the same was as buzzwords like "partnership" or "sustainability" (Utting 2005, p. 375).

During the 1990s, TNCs began to invest in the development of numerous voluntary instruments of social responsibility thereby responding to the public pressure for stronger regulation. The establishment of non-profit organizations which would advocate for the voluntary approach was equally part of TNCs' strategy as the close cooperation with business schools (Shamir 2011, p. 324). By the 2000s, a complex CSR industry (McBarnet 2009) with diverse fields of action had emerged. The set of actors engaging in CSR activities became ever broader: "old" actors such as governments, NGOs, TNCs, international bodies, and global financial institutions in the establishment of a transnational CSR regime now work together with "new" actors such as consultancies and experts, accountancy firms or accreditation, and standard-setting organizations as well as social and environmental auditing firms (Shamir 2011: p. 325). Shamir (ibid.) highlights the particular role of what he calls "market-oriented NGOs", for bringing CSR into practice. These organizations train corporate members on the know-how of CSR, organize conferences, and campaign for the voluntary approach of CSR. In our case study presented below, a market-oriented NGO will also play a key role in the establishment of the studied project. As has been discussed above, the loose definition of the term CSR is the main ingredient of its success, since it allows stakeholders from different and sometimes opposing groups to use the concept for their respective objectives. Hence, a wide range of activities is associated to CSR today. There is a multitude of voluntary initiatives coming directly from the private sector itself; either developed by individual firms, business associations, or public-private multi-stakeholder initiatives. Company or value chain internal codes of conduct, industry standards, certification, and labelling schemes are typical examples of this. A number of non-binding international principles and guidelines have emerged with the aim to orientate corporations on how to assume social, economic, and environmental responsibilities in their supply chains. These soft law instruments are directed to the private sector but also seek to inform regional and national policy processes which in turn are supposed to guide private sector behavior at national levels. Often referred to as the "core set of internationally recognized principles and guidelines regarding Corporate Social Responsibility" (Theuws and van Huijstee 2013) the most important intergovernmental CSR standards today are the UN Guiding Principles on Business and Human Rights (2011) and the Global Compact's 10 Principles (1999), the ILO Declaration on Fundamental Principles and Rights at Work (1998) and its core conventions, as well as the revised edition of the OECD Guidelines for Multinational Enterprises (2012). With the exception of the ILO conventions which are the only legally binding treaties for responsible business conduct, these initiatives only function as soft-law instruments which are meant to provide normative guidance and reputational incentives to business. Proponents of this non-bindingness see the main effect and authority of these tools in their normative influence on the design of other tools which then would trigger down to national policy processes. Indeed, for instance Germany's CSR policy is mainly based on the EU's CSR policy framework which in turn is aligned with shifts in the international interpretation of CSR, as set in the UN Guiding Principles and the OECD Guidelines (Federal Ministry of Labour and Social Affairs n.d.). But there is an ongoing process which might lead to the approval of an UN Binding Treaty for business responsibilities. In 2014, Ecuador and South Africa proposed a treaty to end corporate impunity to the Human Rights Council (Wetzels 2019). Subsequently, the UN Human Rights Council established an Open Ended Inter-governmental Working Group (OEIGWG). The Working Group is mandated to develop the content of the treaty which has as major objective to complement the voluntary frameworks on CSR (Global Interparliamentary Network n.d.). The OEIGWG's draft of the "Legally Binding Instrument to Regulate, in International Human Rights Law, the activities of Transnational Corporations and other Business Enterprises" (2018) sets out how national states should implement and enforce the regulation of business activities and ensure that business actors respect Human Rights in their domestic legislation (Global Interparliamentary Network n.d.).

Next to the international and respective national treaties on CSR, CSR became also institutionalized in numerous private sector and multi-stakeholder initiatives. The following gives an overview over existing tools and approaches that structure the conduct of CSR in the industrial governance context.

As set out above, there is a multitude of purely private or multi-stakeholder initiatives (MSI) and tools of CSR. Multi-stakeholder forums to foster public-private dialogue on policy development, various reporting and disclosure MSI-organizations, standards, certification and labelling schemes, are among the most important CSR forms which go beyond the "traditional" CSR in the form of philanthropic community support (Utting 2005). While the role and legitimacy of MSI and private standards in international norm-building processes and global value chain governance is still largely discussed on a theoretical basis, in practice they have already become widely influential governmental tools. In almost every sector, MSI-platforms are the realm where important policy decisions are made and a broad set of codes of conduct, standards and other tools to guide the behavior of companies and their supply firms are in place. As a result, the 21st century global governance architecture is increasingly complex and fragmented. Intergovernmental institutions, multi-stakeholder platforms, and purely private associations as well as norms and standards often co-exist independently from each other but target the same policy field what makes it challenging to oversee and streamline these governance tools.

Standards are among the most widely applied mechanisms of TNCs in this context. They mainly fulfil two main governmental functions. They give technical information about production processes and product requirements which improves industrial governance and performance. They also provide guidance to the consumers which help them make consumption decisions. Reardon at al. identify standards as "credence goods" (Reardon et al. 2001). An influential definition of standards was given by (Nadvi 2008, p. 325) who defines standards as a "…commonly accepted benchmarks that transmit information to customers and end-users about the product's technical specifications, its compliance with health and safety criteria or the processes by which it has been produced and sourced."

The particular role CSR in the form of standards plays in global value chains and their governance, especially at the producer level, is the topic of this study.

The rapid and uncoordinated spread of many different types of standards within industry and in so many different institutional settings led to a real jumble in the field of transnational governance through standards. Already in 2002, Nadvi and Wältring made an attempt to categorize standard systems according to their content and actors involved. They make a major distinction between product and process standards (for this see also Henson and Reardon 2005). Product standards are mostly of a sector-wide nature and provide technical information to which suppliers have to conform. Product standards have a long history and go back to the end of the 19th century (Kaplinsky 2010, p. 3). The first international food quality standards, for instance, were established in 1962 by the FAO and WHO in the form of the Codex Alimentarius (Trienekens and Zuurbier 2008). Product quality standards are directed towards suppliers and are meant to reduce information complexity between different production segments (cf. Sturgeon 2002). However, during the 1980s, as Nadvi and Wältring carved out, there was a gradual shift from product standards to the use of process standards. The latter mainly refers to management practices concerning the production process (Nadvi and Wältring 2002, p. 6). The growing importance of process standards can be attributed to the increase in consumers demanding transparency and justice in global value chains. The call for evidence on TNCs' willingness to behave more socially responsible is therefore closely linked to the rapid development of many social and environmental process standards. Though, this distinction between the two forms is hazy in many cases, as the quality of a product is often a result of its production process and both forms may comprise many different technical and management aspects of production processes and end product. Nadvi and Wältring (2002) went on to distinguish two different forms of process standards: quality management standards as well as environmental and social standards.

Environmental and social standards have been broadly discussed as a response of business to increasing consumer groups' campaigns for ethical, social, and environmental concerns of production (Clapp and Fuchs 2009; Ellis and Keane 2008; Hatanaka and Busch 2008; Jaffee 2012; Nelson and Tallontire 2014). In the need to convincingly respond to civil society's concerns, TNCs included environmental and social standards as core management elements of industry competition. Thus, there has been a rapid increase in the use of standards and related tools such as codes, labels, and certification schemes with a trend towards their development in public-private networks. The cooperation between business and public entities for the development of standards is particularly often done in global value chains which are environmental resource-intensive as well as labor intensive as it is the case in the mining, agriculture, forestry, and textiles sectors (Nadvi and Wältring 2002, p. 22 ff.). While CSR by its nature is concerned with social and environmental factors of production processes, the line between quality and social/environmental standards became more blurred in the past years and also quality features became part of CSR—as will be shown with the presented case study.

Table 2.1 provides an overview of the two types and the different generations of standards as classified by Nadvi and Wältring (2002) with a special emphasis on the food and agriculture sector, applied by the author. In order to get a systematic picture on existing types and generations of standards, the table provides examples of common standards in the agriculture and food sectors with their respective actors involved and contents.

Among the most influential, non-sector-specific multi-stakeholder developed policy instruments, for instance, is the ISO 26000 Guidance on Social Responsibility. It seeks to provide "guidance on how businesses and organizations can operate in a socially responsible way" (International Organization for Standardization 2018). The standard sets out seven core aspects of social responsibility (organizational governance, human rights, labor practices, the environment, fair operating practices, consumer issues, and community involvement and development). However, the voluntary standard is only meant to provide guidance, there is no corresponding certification scheme for it in place. Sustainability certifications became a major strand of CSR over the past ten years and thereby gained an important position in the global regulatory architecture. Many companies engaging in global value chains now use multi-stakeholder certification systems and labels for the communication of their social and environmental efforts to consumers. Two broadly recognized standards, both developed on multi-stakeholder basis, are the Global Reporting Initiative (GRI) and the accountability standard SA 8000. Whereas the former develops sustainability principles which seek to guide sustainability reporting efforts themselves, the latter provides concrete indicators which allows for businesses to be certified by the organization and is currently the leading social certification standard for factories and organizations (Social Accountability Accreditation Services 2018). Common agriculture and food specific standards with global reach are GlobalGAP, Codex Alimentarius, Forest Stewardship Council, Fair Trade Labelling Organizations International, Ethical Trading Initiative, Rainforest Alliance/SAN Sustainable Agriculture Standard, and the European Initiative for Sustainable Development in Agriculture (EISA) (Genier et al., p. 31). Established product- and crop specific standards are provided by certificate organizations like Rainforest Alliance (including the former UTZ Certified), Common Code for Coffee Community, the Marine Stewardship Council, FairWild Standard, Global Aquaculture Alliance etc. (Genier et al.). Most of these schemes are developed on a multi-stakeholder basis but a number of them in a completely private setting (eg. GlobalGAP).

Table 2.1 Selection of process and product standards in agriculture

Type	Generation	Examples	Actors involved	Content
Quality management	1st: generic	ISO 9000	Private sector and national standardization bodies	Quality management principles
	2nd: Sector-specific	GlobalGAP	Private sector	Food quality and crop management
	3rd Company based	Carrefour Meat Quality and Safety Seal	Lead firms in value chains	Food quality
Environmental and social standards	1st: Company codes of conduct	Mars Supplier Code of Conduct	TNCs and their suppliers	Self-obligations at firm and supplier level, internal formulation and implementation
	2nd: business defined sector codes and labels	Ethical Trading Initiative (ETI)	Business associations	Sector-specific codes and labels formulated and implemented by business associations
	3rd: business defined international standards	ISO 14000	Business, national standard bodies	Environmental management standards
	4th: Business & NGO defined sector-specific codes and labels	FSC, RSPO, MSC, UTZ, RA	NGOs, civil society organizations, unions, business	Sector-specific codes and labels implemented mainly through NGO-business partnerships
	5th: Tripartite defined generic social standards	SA 8000, ETI	NGOs, business, certification bodies, governments	Tripartite social minimum standards

Source: own elaboration based on Nadvi and Wältring 2002.

This introduction to common standards in the food and agriculture sectors should provide the reader with a background on the complexity of the governance structure since these schemes are important tools for the implementation and documentation of TNCs' CSR efforts. Where TNCs initially limited their CSR activities to the conduction of philanthropic flagship projects such as the building of schools, hospitals, or other social infrastructure, the shift towards the usage of standards implies a drastic increase in their fields of action. At the same time, the expectations towards TNCs have also been amplified. For instance, the corporate due diligence concept as an important part of the UN Guiding Principles on Business and Human Rights or the EU strategy for CSR calls for TNCs' extended engagement with all different groups of value chain stakeholders and the larger institutional environment along their whole value chain in order to assure the desired social and environmental behavior of suppliers. In practice, this is mainly expected to be achieved through the development and enforcement of standards and the disclosure of the results through monitoring and certification schemes (UNCTAD 2012). These forms of CSR go far beyond the initial focus of CSR to avoid negative impacts on society and instead require an increase in implementation and regulatory activities. Thus, enforcement capacities at local levels, i.e. a more proactive value chain governance of lead firms. Furthermore, resulting from this trend, TNCs are actively participating in all aspects of value chain governance: They participate in the development of the standards, in their implementation at the various value chain levels, and in the monitoring and reporting of the implementation process and outcome. They do thereby assume a strong position in the global governance constellation. As Tallontire (2007) highlights, TNCs are the driving players at all three levels of power division—the development of rules, oversight, and implementation—the classical separation of powers in this setting is actually dismantled.

2.4   Concluding Thoughts on CSR as a Developmental Tool

As has been discussed above, due to their increased position and expanded field of action in the global setting, TNCs have been called on to assume broader responsibilities in global governance challenges often with a link to local development processes. While some authors treat TNCs akin to citizens and see their role in the active participation in public deliberative processes, other CSR proponents favor the new role of TNCs as important developmental agents who are in the best position to provide public services. Another group of CSR supporters limits its demands of TNCs to take measures which go beyond the traditional duties to prevent human rights abuses. Instead this group expects TNCs to implement measures to prove their steps towards the reduction of misbehaviors in their value chains. Acknowledging the fact that theoretical debate on the scope and extent of businesses' social responsibilities is actively ongoing, there seems to be a general consent that responsibilities should no longer be limited to the core of business activities, that is the economic field, but widely enter other societal spheres.

While some authors recognize CSR as the third way of international development between hard regulation and market-oriented development (cf. Utting and Marques 2013), others point to the fact that TNCs are first and foremost economic entities which engage in development countries for commercial and not for philanthropic reasons (Lenssen and Blowfield 2005). Moreover, the generalized euphoric acknowledgement of CSR as a means to foster local development is blind regarding underlying relationships between prevalent economic structures and social exclusion, inequality, and poverty (Newell and Frynas 2007; Frynas 2005). CSR-critical scholars argue that many instruments of CSR, such as standards and codes which are mainly directed to suppliers in the Global South, are developed in the Global North; too far from the target areas. Therefore, they might not be responding efficiently to the real local challenges (Blowfield 2007). But at the same time, they also appear as a paternalistic governmental tool, since the targeted groups would have to align their activities with them without being represented in the elaboration of their content.

As it stands, there is little empirical knowledge on the implementation and actual impacts of private standards at the local level. Many studies on companies' benefits of CSR have been conducted but little is known about local perceptions and effects on the intended beneficiaries (Blowfield and Murray 2019). Even if many CSR theories suggest that CSR is part of a highly deliberative processes within global governance, in practice the public participation in debate on the content of standards and, importantly, on which ethics and which ideologies are finally implemented by CSR, is limited (Utting 2007). Since standards are one major tool for value chain governance, they cannot be estimated as neutral development instruments at the same time (Blowfield and Frynas 2005). Furthermore, and importantly, being invited to sit on a negotiating table does not necessarily translate in an equilibrated bargaining process. Many civil society representatives are far less experienced in active participation in such high-level deliberations compared to professionals and experts representing TNCs or Western governments (Cheyns and Riisgaard 2014).

Yet, the implementation of private sustainability standards and their corresponding certification schemes became a substantial part of TNCs' CSR strategies and, as set out in the guiding hypotheses, likely has not only effects on the targeted producers, but also more far-reaching ones in the respective local sector constellations. Therefore, this chapter sought to foster a better understanding what CSR today is, but also where it came from and how it was shaped strategically and politically over the past decades. This opens the view for a broader analysis of CSR as a strategic governance tool in a given global supply chain. How such governmental aspirations of TNCs can be theoretically captured and described is content of the next chapter.

Consent and Control: A Conceptual Framework for the Study of CSR in the Global Cocoa-Chocolate Chain

3

CSR is a multi-faceted umbrella concept and a large number of studies approach it from a wide variety of angles. While CSR's justification, content, and reach are under continuous debate, its application in practice and its establishment in international agendas and treaties are diverse and pluralistic (see chapter 2). The present study is particularly interested in understanding how CSR is applied in the Global Cocoa Chocolate Chain (GCCC) and how it shapes the institutional environments, especially at the local level of production. As set out with the research questions in the introductory chapter, the interest lies in understanding if, and if so, how CSR helps TNCs to increase their control and influence over local production. If CSR would show to help TNCs to increase control in the GCCC, it should not any longer only be considered as a prominent public relations strategy or a philanthropic means of positive duties but, and this is the key argument of the present study, as a governance tool that powerful value chain actors apply in their chains to respond to multiple sustainability pressures.

As has been already indicated in the introduction and will gain further attention in chapter 4, the GCCC faces a number of sustainability challenges that TNCs in the GCCC need to address. From an industry perspective the challenges can be summarized with the notion of "double pressure", categorizing challenges in the GCCC into two main groups. One pressure is the perceptions and attitudes of key stakeholders in the GCCC, such as producers and consumers. Over the past two decades, the awareness on some grievances in the chain has increased. Consumer groups in particular have urged the industry to act. At the same time, there are a rapidly increasing number of sustainability challenges at the production level which jeopardize a reliable flow of the cocoa beans in the future to meet the growing global demand. TNCs need to find efficient responses to this

F. Ollendorf, *The Transformative Potential of Corporate Social Responsibility in the Global Cocoa-Chocolate Chain*, (Re-)konstruktionen – Internationale und Globale Studien, https://doi.org/10.1007/978-3-658-43668-1_3

double pressure which would appropriately manage the two aspects in order to keep the system in status quo.

From its very nature, CSR has the ability to respond to both of these of pressures. CSR is directed to the producers and their production capacities and serves as evidence of change to satisfy consumers. Additionally, the implementation of CSR strategies itself shapes institutional environments at the respective levels of action and can be applied for that purpose. It is therefore worthwhile to conceptualize CSR as a governance strategy which helps TNCs in the GCCC to respond to this double pressure. Given the double pressure effect, a differentiated framework that allows the systematic analysis of CSR as a tool for improved governance is needed. But what is meant with governance in this context? Governance became a key bridging concept in many social science disciplines and highlights the link between actors and institutional levels (Hartmann and Offe op. 2011). The present study takes the basis from the very meaning of governance which refers to steering, in other words the coordination of actions and the development of regulatory structures that frame problem-solving processes. Mayntz (2004) provided a broadly recognized definition of governance as the

> "total of all coexisting forms of collective regulation of social issues, from institutionalized civil society self-regulation and various forms of cooperation between state and private actors to sovereign action by state actors." (Mayntz 2004, author's translation)

Two key features of her concept are the regulatory structure and its intentionality. The regulatory structure refers to task-related institutional arrangements such as regulatory authorities, norms, and governance instruments. What is primarily understood as intentionality here is the regulation of collective issues that shape particular problem-situations while justifying the regulation with the collective interests of the given group (Hartmann and Offe op. 2011), hence the assumption to pursue a common goal.

In this chapter, the conceptual framework for the study of CSR as a governance tool will be developed which takes into consideration the described double face of CSR. The framework benefits from the combination of two governance theories which apply the notion of governance in very different ways to the global realm and therefore allow to analyze both the regulatory structure and the intentionality: global value chain governance (which looks more into the regulatory, industrial governance aspects) and a neo-Gramscian view on global governance (which sheds more light on actors' intentionality). The Global Value Chain approach allows for systematically describing the composition of a given value chain and understanding the mechanisms of industrial governance and its particular organizational aspects of inter-firm relations. While a neo-Gramscian perspective can shed light on governmental activities which seek to shape perceptions of people and create an enabling, chain-embedding institutional environment. The integration of these two approaches into the consent and control framework comprehensively responds to the assumed two-sided direction of CSR governance. Figure 3.1 visualizes the key components of the governance framework developed in this study.

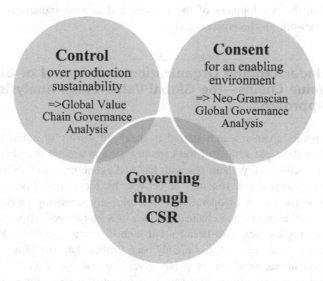

Figure 3.1 Conceptual framework: control and consent—governing through CSR. (Source: own elaboration)

In this chapter, the conceptual framework of the study is elaborated by dedicating one subchapter to each of the two theoretical approaches combined in the framework. Subchapter 3.1 introduces to Global Value Chain Analysis (GVCA) as a methodological tool for the assessment of global value chain functioning. Particularly its axis 'governance' will be discussed in more detail with application to current research streams and critiques. This provides the background for a general understanding of global value chains and the following description of the GCCC and the role of CSR in it (chapter 4). With regard to the governance of a chain, however, shortcomings of GVCA are presented and discussed. After briefly highlighting the ideological journey of commodity chain research and presenting the core theoretical features of the approach, the subchapter closes by pointing out key limitations of the approach. In subchapter 3.2, in a first step, an introduction to the theoretical basis of neo-Gramscian works is given and Antonio Gramsci's concept of a cultural hegemony through consent and coercion presented. In a second step, the subchapter introduces to contemporary receptions of Gramsci's thought and to their further elaborations in the field of International Political Economy. In the third part of 2.2, the neo-Gramscian perspective is employed to CSR and its transformative effects in global value chains. Finally, in subchapter 3.3, the nexus between the two approaches will be unfolded which culminates in the development of the consent and control framework as core guiding framework for the present study.

3.1   Understanding Organization and Control of Global Value Chains—the Global Value Chain Analysis Approach

Early commodity chain research started in the 1970 s with an initial focus on understanding the rapid spread of the capitalist system around the globe. Since its early days, commodity chain research has undergone profound shifts in theoretical underpinnings and research objectives. Its roots stem from a critical assessment of the modes of operation of the globally spreading capitalist system through global production chains. The idea of a global production chain was picked up during the next two decades but with different objectives. From the 1990 s onwards, chain research has mainly been further elaborated from an organizational and management theory perspective. The concept of GVCA became the most widely accepted today. What made it attractive for researchers from different disciplines investigating global production flows is its analytical set that is concerted in four axes which together provide a comprehensive picture of a given chain: The axes "input-output structure, geographic scope, institutional environment and governance", out of which the first two are largely descriptive while the governance and institutional axes are more of explanatory nature (Sturgeon 2009b). Till today, the main theoretical debate concentrates on the governmental axis while, as some chain researchers detect, the institutional axis is receiving only little attention (Bair 2005, p. 154; Fold and Neilson 2016).

3.1.1   The Evolution of Global Value Chain Analysis—the Ideological Journey of a Research Concept

Jennifer Bair (2005, 2008) draws a genealogy of the global value chain theory and identifies a sequence of three main concepts, one building on the former but differing regarding the theoretical foundations. First centered on the analytical term "Commodity Chains," then to the analysis of inter-firm networks organization in global industries under the term "Global Commodity Chains"[1], and finally to the current dominate analysis of sectoral logics of global industries from an international economy perspective under the term "global value chain analysis" (Bair 2005, p. 160). In addition, Fischer et al. (2010) and Knudsen and Fold (2011) also underline the importance of a fourth stream, the "Global Production Network" approach. Particularly applied by economic geographers, this framework examines linkages between all stakeholders within a Global Production Network at a national and international level. However, since the chain metaphor emphasizes a hierarchical constellation among stakeholders, it is more appropriate for the present study which puts particular emphasis on TNCs' value chain interventions in the course of their CSR programs.

According to Bair and other scholars (e.g. Talbot 2010), the roots of chain research lie in World System Theory (WST). In 1977, Hopkins and Wallerstein introduced the term "commodity chain" (Hopkins and Wallerstein 1977) in order to reframe the scope of global capitalisms' territoriality. The concept was developed to enable the describing of the new labor and production processes leading to the increasing core, semi-periphery and periphery distinction and the unequal distribution of awards in the global economy. In the following decade, commodity chain research mainly concentrated on understanding the process of capital accumulation, the distributional effects between the different production segments as well as the monopolization of the most profitable production segments by the dominant actors in a given commodity chain (Bair 2005).

Among many other concepts which start from WST's commodity chains, Gereffi's framework on global commodity chains, published in 1994 in the collective volume "Commodity Chains and Global Capitalism" (Gereffi and Korzeniewicz 1994), became widely influential. His concept constituted a major disjuncture with the former approach of the WST (Bair 2005, p. 155, 2009, 14ff). The new focus of interest was on "understanding the organizational dynamics of global

---

[1] In the francophone literature, the filière concept provides a similar framework, Raikes et al. 2000.

capitalism" (Gereffi 1996b, p. 427). With this, the organizational aspect became more important and outstripped the system theory-based one. The main focus was now twofold. First, global commodity chain research is more concerned with depicting patterns of lead firms' control over their suppliers. Second, based on the understanding of these key linkages, best strategies for exporting-firms from the "developing world" to benefit from their integration in the global commodity chain are to be identified. This new focus could be interpreted as a shift from the goal of critically assessing global capitalism's mechanisms towards the main ambition to detect the best form of integration of developing countries' firms into the capitalistic system in order to improve their status in it. This process in the literature is described with the term "value chain upgrading".

Hence, the focus of Gereffi's Global Commodity Chain approach (Gereffi 1996a) is primarily on inter-firm relations in a given commodity chain. In contrast to the WST scholars' conception of commodity chains as a necessary historical continuity of the global capitalist system, Gereffi's main interest lies in "the fact that international production and trade are increasingly organized by industrial and commercial firms involved in strategic decision making and economic networks at the global level" (Gereffi 1995, p. 113). This condition led Gereffi to theorize the broadly received distinction between two prevailing forms of commodity chains—they would be either producer-driven or buyer-driven. Accordingly, a producer-driven chain would be characterized by a high degree of control over backward and forward linkages in the production system exercised by transnational corporations or large integrated industrial enterprises. In such chains, TNCs typically would subcontract components of most of the labor-intense production processes internationally but exercise administrative control in their headquarters. Gereffi ties the emergence of this form of chain governance to the rise of TNCs in the 1960 s (Gereffi 1995, p. 115).

In addition to the producer—buyer-driven divide, another important aspect of the commodity chain analysis is the concept of upgrading. The main idea here is the optimization of the benefits stemming from Global South firms' integration in global commodity chains. The rationale is the idea of a trickle-down effect. Meaning, the assumption that the industrial upgrading of a firm would trickle down to the national economy, and therefore contribute to improve the position of a nation-state in the global economy. A high amount of potential to boost a developing nation out of poverty is thus attributed to its integration into global commodity chains (Talbot 2002). Global commodity chain researchers therefore often ask how the governmental structure of a value chain—mainly understood as organizational structure of chain management—affects the opportunities of exporting firms, particularly from developing countries. Thus, the focus is on how to enter the chain, how to gain access to skills, competences, and supporting services needed for a successful integration in a global chain, and what potential of industrial upgrading exists for participating firms and industries as well as for the societies they operate in (Gereffi et al. 2001b).

The fundamental disruption with the WST becomes most visible here. Whereas the WST identifies a "developmentalist illusion" (Arrighi 1990) which is due to the capitalist system's intrinsic reproduction of periphery and center structures within the setting of a global commodity chain (Bair 2005, p. 157), the Global Commodity Chain approach is optimistic regarding upgrading and improvement through the integration in the chain. Of course, Gereffi and colleagues acknowledge the huge asymmetries in power relations in the chains and one of the main objectives is to describe how they are organized and maintained by lead firms. But they step away from the critical assessment of structural reproduction of social systems and follow a rather pragmatic search for the best form of firms' participation in this system. This new focus on the firm level itself rather than on the systemic level constitutes a major shift from a macro level perspective to a micro level of interfirm relations and particularly upgrading strategies (Bair 2005, p. 154).

The continuous increase of the fragmentation of global production networks across "geographic space and between firms" (Gereffi et al. 2005, p. 80) and the differences in success stories regarding industrial upgrading motivated chain researchers to investigate the multitude of coordination activities in these new inter-firm relations (Gereffi et al. 2001a). In 2005, Gereffi, Humphrey and Sturgeon came up with their latest and most widely received model of commodity chain methodology, the Global Value Chain Analysis approach (GVCA). The main revisions of the former global commodity chain approach concern the governmental axis. This concept provides a new typology of chain governance, thereby overcoming the producer—buyer-driven chain dichotomy, which still plays an important role in today's chain research (Ponte and Gibbon 2005, p. 3). The new typology embraces five forms of chain governance, locating market-based relationships between firms at the one end and vertically integrated firms at the other end of the coordination spectrum (Gereffi et al. 2005, p. 83). The role of industrial upgrading, particularly by developing countries' firms, remains crucial and strongly influenced by transaction costs and rent theory (Bair 2005, p. 160, 2009, p. 12).

The genealogy drawn by Bair made the intellectual shift of the chain research visible. Initially inspired by dependency theory and structuralist development economics, the most recent work mainly draws on theories of international business, industrial organization or trade economics (Bair 2005, p. 160). Hence, the objective of much of todays' chain research widely differs from the initial objective of the WST scholars and is not a critical analysis of the social reproduction of global production systems anymore. The analysis of inter-firm relationships and upgrading strategies is at the center of much of global value chain research (Ponte 2019; Humphrey 2007; Aoudji et al. 2017; Kilelu et al. 2017; Armando et al. 2016). This more pragmatic approach might be more applicable and suitable to give concrete policy advice to development practitioners. However, this shift to a focus on organization and coordination patterns lead to a technical interpretation of development rather (cf. Charlery de la Masselière, Bernard 2002) than to an understanding of development as a societal and political process.

Furthermore, as will be discussed in the following, this tendency of technization of development as an economic upgrading process and the concentration on mere inter-firm relations limits our understanding of the complexity of lead firms' activities which aim to keep or to extend control in their value chain. Some influential GVC scholars, like Ponte and Sturgeon, consider the discussion of TNCs' relation with society and governance institutions as not being part of GVC research. They argue that the diverse forms of influence of TNCs on "governments and international organizations to obtain favorable rules" (Ponte and Sturgeon 2014, p. 200) is not part of the GVC study and remains task of the neighboring disciplines, particularly to various strands of political economy (Ponte and Sturgeon 2014, p. 200). However, from the evolutionary perspective on GVCA summarized here, there seems to be no convincing reason for such a conclusion. First of all, GVCA roots lie in exactly such ambitions to uncover societal relations that reproduce capitalist systems of power relations, and, secondly, the GVCA approach can be read as a dynamic and open methodological concept which has been introduced, further developed, and applied by scholars from different disciplinary backgrounds with distinct research interests regarding global value chain dynamics. This eclecticism is actually highlighted and appreciated earlier by Ponte himself in an introductory paper to the special issue "Governing global value chains" in "Economy and Society" that he edited jointly with Bair and Gibbon (Gibbon et al. 2008, p. 334). Still, there seems to be the need to demonstrate the necessity of a value chain governance understanding which considers struggles over power distribution not only in the direct industrial linkages but also regarding enabling and embedding institutional environments and even regarding stakeholders' perceptions and mindsets.

3.1.2   Global Value Chain Analysis as a Methodical Framework

Since their introduction to the global commodity chain concept, the four analytical axes remain the core framework of global value chain analysis. Gereffi defines these axes as follows:

> "a value-added chain of products, services, and resources linked together across a range of relevant industries; a geographic dispersion of production and marketing networks at the national, regional and global levels, comprised of enterprises of different size and types; a governance structure of authority and power relationships between firms that determine how financial material, and human resources are allocated and flow within a chain; and an institutional framework that identifies how local, national and international conditions and policies shape the globalization process at each stage in the chain." (Gereffi 1995, p. 113)

Together, these four axes cover central aspects of a chain configuration and help to understand important structures of global value chains. In the following, based on Gereffi and Fernandez-Stark's primer (Fernandez-Stark and Gereffi 2011) on Global Value Chain Analysis, a short summary of the practical handling of how to approach a complex production network with the help of the four axes will be given.

Axis Input-Output Structure
In the first step of value chain analysis, the main activities and segments in a global value chain are to be identified. Services and goods introduced at different production stages will be mapped according to the value-addition to the product. Thereby, the returns netted for participating chain actors become visible. In the second step, the characteristics and dynamics of each chain segment will be analyzed more in detail. Here, the focus is on defining the type and structure of companies participating in each industry segment.

Axis Geographic Scope
In this part of global value chain analysis, the interest lies in the "identification of the lead firms in each segment of the value chain" (Fernandez-Stark and Gereffi 2011, p. 7) and their geographic positions. Knowing the localization of the lead firms also informs the assessment of the country-level position.

Axis Institutional Context

This axis refers to the institutional embeddedness of a global value chain and aims at the identification of the "local, national and international conditions and policies" (Fernandez-Stark and Gereffi 2011, p. 11) which shape the chain at its different segments. The driving interest of the analysis is the assumption that these aspects would directly affect firms' possibilities of insertion and upgrading in the chain. Thus, the local economic conditions, e.g. the availability of inputs and infrastructure or the accessibility of resources, as well as the social ones, e.g. the skills of the local population, are seen as crucial factors shaping the participation in the chain. Furthermore, national and international regulations and policies, such as tax and labor regulations or subsidies, affect the functioning of a chain. In a global value chain analysis, assessing the institutional context of a chain also implies mapping the stakeholders of the industry (Fernandez-Stark and Gereffi 2011, p. 11).

It is important to recognize here that the notion of institutional embeddedness is only considered in a one-way direction. Institutional environments almost appear here as static, or as a political sphere where private actors would have no access. Coming back to one of the main hypotheses of this study, a central assumption states that CSR is a tool for lead firms to pro-actively participate in decision-making and agenda-settings which shape their supply chain. Thus, narrowing the institutional axis to only one direction where external institutions influence chain conditions bears the risk of ignoring dynamics where lead firms or other chain stakeholders at any given production segment make efforts to shape the institutional environment around them. Importantly, these efforts would also impact existing governmental structures in the chain. The analytical dissociation of these two axes might hamper an appropriate understanding of governance dynamics stemming from pro-active CSR.

Axis Governance

The analysis of the governance structure of a value chain aims to understand the mechanisms of coordination of chain activities by the lead firms and their strategies to control their suppliers in the chain. In the primer, Gereffi and Fernandez-Stark refer back to Gereffi's definition of governance in his initial chapter on global commodity chains where he defines it as follows: "authority and power relationships that determine how financial, material and human resources are allocated and flow within a chain" (Gereffi 1994: p. 97).

As mentioned above, five types of governance structures are established: market, modular, relational, captive and hierarchy governance (Gereffi et al. 2005, pp. 86–87). These forms of GVC governance can be read as a scale where "market governance" at the left end of the scale represents the lightest version of lead firms' control whereas "hierarchical governance" would exert the most control and influence along the chain. Gereffi et al. developed these governance types based on economic theories of transaction costs, production networks as well as technological capability and firm-level learning (Gereffi et al. 2005, p. 81). As already mentioned above, there is no place for the discussion of TNCs' attempts to create an enabling environment in which their suppliers operate in, hence no institution-building efforts can be taken into account with the prevailing GVCA's understanding. The governance axis has been broadly debated in the GVCA literature but very few positions have been identified which support this argumentation.

3.1.3   Limitations of the Industrial Governance Understanding in Global Value Chain Analysis

Global value chain governance by late has been broadly discussed as coordinative decisions of lead firms, mainly by economists and economic geographers. Thereby,

> "concrete practices and organizational forms through which a specific division of labour between lead firms and other economic agents involved in conceptualization, production, and distribution of goods in global industries is established and managed" (Gibbon and Ponte 2008)

have been in the center of attention. This shows a widely apolitical reception of value chain dynamics and therefore, at a first glance, seems to make GVCA useless for other social scientists who are more interested in grasping societal power relations as a whole and not limited to inter-firm relations, as are political scientists or sociologists. However, some GVCA authors argue that governance has more faces than the coordination of inter-firm relations and that, importantly, questions of power have to be considered (Sturgeon 2009a). Gereffi himself in a joint article with Mayer (2010) takes into consideration main factors which influence the effectiveness of private governance: the structure of a particular chain, the extent to which the role of the brand identity also shapes strategic decision, the degree of consumer pressure that has to be handled by the lead firm, and the possibilities with which the commercial interests of the lead form can be aligned with environmental and social concerns. With this, it becomes clear that the interaction of a lead firm with chain external actors and environments is of much greater importance than assumed in the main literature body.

More arguments have been developed to support the multi-dimensionality of chain governance. For instance, Bair (2008) makes the point that the actual social embeddedness of a given chain also depends on existing interpersonal relations within the given chain and highlights the network character of industrial supply chains. Palpacuer (2008) reminds that lead firm decisions importantly are influenced by their shareholders' interests and the financialization strategies of this clientele. Scholars applying convention theory to global value chain governance have argued to approach the field of value chain governance as an attempt of normalization and aligning business practices to multiple competing orders of worth as for instance market, industrial, domestic, civic, inspirational or opinion orders of worth (Ponte and Gibbon 2005). If some orders of worth were competing, it would not be clear which conventions dominate the firms' decisions. Departing from this, the necessity to take the mutual constitution of micro, meso, and macro levels into account and to include regulatory and societal pressures becomes more evident. For example, a civic convention may judge the quality of a product according to its impact on society and environment. The development of standards and certification schemes which concern social and environmental factors as an immediate reply to such external pressures gives way for a more prominent recognition of chain external relations of a lead firm. The role of social and environmental standards in global value chain governance has gained more attention of late, but the pro-active strategy to participate in the development of multi-stakeholder governmental systems, and thereby to shape institutional settings external to the immediate industrial chain, has been widely neglected so far in the GVCA literature.

For the study of CSR's transformative potential in the GCCC, the prevailing limitation of GVCA to an internal governance understanding hinders researcher to gain a more holistic picture of lead firms' activities. As has been suggested in the hypotheses in the introductory chapter, TNCs' CSR interventions are directed towards different dimensions and should be understood as tools for a more pro-active exertion of influence in a global value chain. The companies' need to deliver direct and visible improvements in their value chains which increase their output legitimacy and prevent hard-laws can be deduced from the several pressures TNCs in the GCCC currently face. Simultaneously, structural requirements at the level of cocoa production also need to be handled in order to sustain and extend their dominant position in the chain. This double challenge also requires a more comprehensive analytical framework while assessing and discussing TNCs' CSR activities. With the hypothesis that the main objective of CSR is to sustain the dominant position, assure production flow and increase influence in order to improve governance, the present study recognizes the need for a TNC to establish institutional conditions in the chain that allow the company to exert more control and to gain better access at the local level.

The GVCA approach provides a good starting point for this objective and for approaching CSR in the GCCC. It offers a systematic method to describe a value chain structure, for example the different production segments or the geographic scope of value-added distribution. It is an important tool for the explanation of existing forms of industrial relationships between stakeholders within a given chain. But the current mainstream understanding of global value chain governance limits itself to the explanation of inter-firm relationships and understands value chain governance as managerial decision-making regarding these relations. This is definitely a central part for the study of CSR's transformational potential in the GCCC. However, as it is argued here, CSR is not only directed to manage chain internal relationships but has a broader governance ambition. As argued above, one of its main objectives is the interference into production processes in the chain with the stated aim of achieving more sustainable practices. Initially, in the GCCC CSR's content centered on the elimination of the worst forms of child labor—a form of production directly intertwined with the broader institutional environment of the chain. Thus, it was the explicit ambition of CSR to proactively transform the chain's embedding institutional conditions in order to achieve the respective goals. With the predominant limitation of GVCA to chain internal industrial factors, these dynamics remain excluded from the analysis and important efforts of a lead firm to improve control along the chain cannot be captured. Therefore, a combination of GVCA with a global governance concept of hegemonic power seems a more suitable approach for the present study.

3.2   CSR as a Consent Strategy—Neo-Gramscian Perspectives on Global Governance

Recently, researchers within the value chain perspective have argued for an understanding of chain governance which goes beyond assessments of inter-firm relations and organizational patterns in a chain. Based on Levy's work on "Political Contestation in Global Production Networks" (2008), Bair and Palpacuer (2015) suggest combining the prevailing understanding of global value chain governance in terms of industrial or chain internal governance with a neo-Gramscian inspired concept of global governance which also considers the process of the creation of norms and rules regarding global production. For the present study of

CSR's transformative effects in the GCCC, the combination of industrial governance, that is "the coordination of relationships among actors in a global value chain" (Bair and Palpacuer 2015, p. 1), and a concept of global governance is very promising. Particularly a neo-Gramscian perspective on global governance appears to be fitting to the objectives of the study. This perspective, originating in the field of International Political Economy (IPE), allows one to approach CSR as a hegemonic tool for improved value chain governance which takes into account the pluri-centered exercise of power followed by lead firms in a given production context. The combination of these two different governance understandings led to the development of an industrial-global governance nexus for CSR.

3.2.1 Antonio Gramsci's Concept of Cultural Hegemony

Antonio Gramsci was an Italian intellectual, journalist, politician, philosopher, and one of the most influential early neo-Marxist thinkers. With his concept of hegemony, he has given major impulses to writings in political theory, theory of the state or political philosophy, and other disciplines some decades after his early death in 1937. The "originality" (Buckel and Fischer-Lescano 2007, p. 11) of Gramsci's work lies in his extension of the prevailing understanding of political leadership, which until then was seen as being mainly guaranteed through coercive means. Further elaborating on Marxist theory and criticizing its economic basis, Gramsci introduced a pluri-centered conception of power which draws on the interplay of material, institutional, political, cultural, and ideologically-discursive elements of leadership (Bieling 2016). For some authors, this enlarged approach to power made the traditional, one-sided approach to state power even obsolete (Hall 1991).

Reading Gramsci's work, it is particularly important to take into account the context of the historical period of his writings. His thoughts were marked by a profound historic double change which he sought to explain: the defeat of the workers' movement and the rise of fascism in Italy as well as the stagnation of the Russian socialist movement after the October revolution and Lenin's death. At the same time, in Italy and many other Western European countries, Fordism and Americanism expanded rapidly what led to a profound restructuring of production and of many other domains of life (Candeias 2008, p. 17). Being actively involved in the fight against Italian fascism in the early 20th century, the main motivation for Gramsci was to be able to articulate an informed critique on domination and to develop effective liberation strategies (Merkens and Diaz 2008, pp. 7–14). In that vein, Gramsci sought to understand the functioning of the rising capitalist society and what it needs for building up alternative forms of state and society (Cox 1983, p. 162) Hans-Jürgen Bieling (Bieling 2016) classifies Gramsci's political and theoretical engagement into three periods: First, the period of political journalism until 1920 when Gramsci published political-strategic as well as culture- and ideology-theoretical essays, second, the period of party-political engagement in Italy's Communist Party and the active fight against fascism which took center stage of his activities after Mussolini's assumption of power in 1922, and third, the period of imprisonment in 1926 as a result of the dictatorship's persecution of political opponents. In poor health, Gramsci was paroled after eight years in prison and passed away in hospital due to a cerebral bleeding at the age of 46 (ibid.).

Despite his severe health problems, during the last years in prison, Gramsci was able to develop his main body of work, the Prisons Notebooks. Mario Candeias (2008, pp. 15–32) highlights that Gramsci did not plan to publish the notebooks in the form he wrote them. They are rather the product of his process of critical knowledge production, and thus of a fragmented nature instead of being a complete and closed piece of theory. Regardless of the format, the Prison Notebooks constitute an innovative analysis of contemporary social challenges in the early 20th century. Gramsci shifted the focus towards new questions such as the character and the different forms of leadership and domination, the meaning of the popular national culture, the development and the state of the civil society and ongoing shifts in societal power balances (Bieling 2016, p. 443). Thereby, the contradictory processes of the normative integration of social domination are at the center of his interest (Bieling 2016, p. 444). For Gramsci, in order to achieve the dominant position in a given society, the power-aspiring social group would have to establish relatively stable societal consent. It is through the sensitive organization of this consent, that political and ideological leadership can be achieved (Simon 1991). Gramsci sees in the achievement of such a relatively stable constellation of societal domination the realization of hegemony, which, however, is to be seen as a dynamic project always challenged by subaltern groups (Jones 2006). Gramsci's conception of hegemony is based on two main strands of thought: The first, Lenin's concept of hegemony, and the second, Machiavelli's theory of the state. Lenin's idea of hegemony referred to a strategy where the proletariat should become the dominant class and give direction through leadership of the allied classes, like the peasantry. Gramsci developed this idea of a proletarian hegemony further and used it to explain the dominance of the capitalist system (Cox 1983, p. 163). The second inspiration, Machiavelli's image of a centaur, brought emphasis to the double face of power, that is the combination of consent and coercion (Cox 1983, p. 164). Instead of continuing to focus on the coercive elements of power of the state, Gramsci put a particular emphasis on the part of consent. Moreover, it is one core aspect of his thought that a stable power constellation is only to be considered as hegemonic when subordinate groups follow the dominant group voluntary and coercion is only latent, thus consent reigns. The major interest lies in the analysis and understanding how the mechanisms of consent are established and maintained and why people accept the leadership of others.

Gramsci describes the process of establishing a hegemonic project as follows: Once a hegemon-aspiring social group has organized itself and articulated its corporate interests, the latter have to be transcended into political claims that are able to achieve universalism and agreements of other social groups. In this process of creating a new societal consent, which is still based on the formerly existing one, the ability of assuming an intellectual and moral leadership is of crucial importance (Bieling 2016; Jones 2006). The dominant group needs to engage with the culture of subaltern groups and incorporate it into the own world view. The key is to shape the national popular ideology in such a way that the interests of the dominant group appear as universal interests of the whole society. Therefore, the dominant group has to engage with subalterns, create alliances, make compromises and concessions, to create the psychological basis that subalterns accept to be dominated. Besides, the dominant group has to ensure the fulfilment of the basic needs of subalterns. That means, to partially share the wealth, and to truly internalize the subalterns' interests in its own agenda—never to such an extent that their own interests would really be affected but as much as needed to stabilize the alliance (Jones 2006, p. 45). Counter-hegemonic philosophies and critiques on the prevailing hegemonic paradigm need to be disarticulated and incorporated in the system. But the challenge remains persistent. Gramsci recognizes the capabilities of the subalterns to emancipate themselves and to become strong protagonists. The hegemonic group therefore has to constantly reply to their demands and to ensure that they have enough to eat, good health care etc. Importantly, the ideology expressed in universal terms and the institutions representing it do not have to appear as those of a particular group and also have to satisfy the demands of the subordinate groups (Cox 1983, p. 168). Hence, next to establishing the economic conditions needed, the hegemonic project also needs to work on cultural, institutional, and legal-political dimensions (Cox 1983; Jones 2006).

Reflecting on the locus of these processes, Gramsci developed the concept of the "integral state" (Gramsci 2012, p. 783). It combines what Marx grasps with his notion of the superstructure, that is the political and the civil society. In this enlarged understanding of the state, the state is seen as a relational construct which consolidates and institutionalizes the prevailing societal power balance and which is always in process of change (Bieling 2011b, p. 12; Candeias 2008, p. 21). The political society comprises public institutions in charge of administrative and juridical tasks and the military which serves as the monopoly on the use of force. In contrast, the civil society constitutes the arena where the actual political struggles over domination take place and where the dominant group has to organize the societal consent. In other words, the diverse discourses and institutions of the civil society are the arenas where the mediation processes between the hegemonic structures and the societal everyday behaviors take place. Hence, the political society being mainly the institutionalized representative of the dominant group, has to penetrate the spheres of the civil society in such a way that civil society internalizes the hegemonic values and meanings and further disseminates them (Bieling 2011a, p. 12; Barfuss and Jehle 2017, p. 108 ff.). For example, civil society organizations, such as the church, academia, the press and, importantly NGOs and advocacy groups (Barfuss and Jehle 2017, p. 108 ff.), are all regarded as organizations of knowledge production. They play a crucial role in the legitimization process of those discourses which shape and create people's mindsets and behaviors (Candeias 2008, p. 22). In a relatively stable hegemonic system, when the dominant group has achieved moral and intellectual leadership, most of the civil society institutions would form their discourses in consistence with the hegemonic social order (Cox 1983, p. 164). Thus, civil society is understood as a complex ensemble of institutions, ideologies, and practices (Candeias 2008, p. 22). Gramsci identifies in civil society the major arena where the struggle over moral and intellectual leadership is taking place and therefore attributes even more political importance to it than to the formal state itself. The "integral state" reflects Gramsci's understanding of state and the nature of power what he has summarized in his often-cited quotation "hegemony armored by coercion" (Gramsci 2012, p. 783).

However, the degree of stability of consent is varying, and often enough, hegemony is threatened by counter-hegemonic movements. In these cases, major adjustments and a reorganization of the system are required in order to safeguard the dominant position. In such moments, these far-reaching modifications in the social and economic structures of the society are established from above and do not rely on the people's participation (Simon 1991, p. 25), thus a "revolution without revolution" (Candeias 2008, p. 17) is taking place which is mainly the restauration of the dominant group's position. Gramsci calls this process a "passive revolution" (Gramsci 2012 GH 1: 102). He describes it as a central way in which the dominant group seeks to keep its hegemonic position alive in times of crises. In this context, Gramsci describes a strategy that he calls "*trasformismo*" which aims to assimilate and co-opt potentially dangerous ideas by undertaking some minor changes in the system and slightly adjusting the policies to them (Cox 1983, p. 165 f.). If all these efforts are successful, a relatively stable power balance is maintained, and the dominant group is also leading the economic realm, Gramsci identifies a constellation that he calls "historic block" (Gramsci 2012 GH 6: H 10.1, §12). A historic block can be understood as a configuration where all elements of hegemony, that is the political, the moral and ideological spheres as well as the economic sphere interact reciprocally and create a larger unity (Cox 1983, p. 167). Including both, dominant and dominated groups in the society, the historic block constitutes a contradictory unity that has to be reproduced constantly and this is what is at the core of Gramsci's understanding of hegemony: Hegemony is regarded as an open societal project that due to its internal contradictions is always contested and transformable by societal practices. Whenever the consent gets too porous, elements of coercion become more important in order to avoid the process of subversion (Candeias 2008, p. 20 f.). Hence, the cultural, economic, and political aspects of hegemony always rely on the latent threat of violence monopolized at the level of political society. However, as Jones (2006) highlights, today, the actual use of violence is of less relevance to keep the internal balance of power in a state. But, he points to a new mechanism of violence that Pierre Bourdieu has conceptualized with the term "symbolic violence". Accordingly, subordinate groups which are developing counter-hegemonic thoughts which are considered to be too threatening to the historic block are sanctioned. For example, by exclusion in political processes, by the omission of their opinions in public discourses, and by other forms of marginalization of outsiders. The hegemonic process can be summarized as successful when the ruling power effectively reproduces its authority, the subaltern groups aspire the values and tastes of the dominant group, and the low status of the dominated groups is reinforced (Jones 2006, p. 52 f.).

3.2.2 Neo-Gramscian Approaches to International Political Economy

Antonio Gramsci's concept on hegemony has influenced many scholars' thinking from a variety of Social Science disciplines including Cultural Studies (e.g. Stuart Hall or Edward Said), Philosophy and Sociology (e.g. Jürgen Habermas or Michel Foucault), and in the Political Sciences particularly in the field of International Political Economy (Robert W. Cox, Stephen R. Gill). This subchapter is limited to an overview of the main neo-Gramscian approaches developed in the field of

International Political Economy (IPE) since these are concerned with questions directly related to the topic of the present study and therefore are most fruitful for the understanding of CSR as tool for improved value chain governance.

Departing from the "classical" strand of International Relations which neo-Gramscian writers have criticized as being trapped in a "methodological nationalism" (Opratko and Prausmüller 2011, p. 13), neo-Gramscian perspectives in IPE have focused on societal power relations between different actors at the transnational level. Thereby, societal interactions which take place in the political and the economic realm as well as the character of contemporary socio-cultural, economic, and political restructuring processes represent the core of analyses (Opratko and Prausmüller 2011, p. 13). Sharing these common points of departure, the neo-Gramscian IPE is not a coherent closed field of theory but rather an ensemble of mutually aligned theoretical fragments (Bieling 2011b, p. 11). Robert W. Cox, a Canadian political scientist, has been the first who transferred Gramsci's work to the global level of power analysis. It was Gramsci's main question, how a social order that systematically disadvantages the majority of subalterns could still persist, that Cox sought to be of equal relevance when analyzing the functioning and global predominance of capitalism. Cox applied the idea of hegemony through consent and the universalization of particular group interests to the international realm. By doing so, he did not look at the theory as a rigid frame but rather as an analytical guide. He complemented it with other approaches and further developed the concept. In this sense, Cox stated that the openness of Gramsci's concept would be an advantage in so far as it makes it applicable to other constellations and questions beyond Gramsci's immediate concerns (Cox 1983, p. 162; Candeias 2008). In order to be unambiguous when elaborating on the concept of hegemony, Cox explicitly broke with the prevailing understanding of the hegemony in the neorealist strand of International Relations. Therein, the term hegemony is used to describe dominant nation states' activities (particularly when analyzing US power) which are considered to establish and enforce rules in geopolitics and international economic relations (Opratko and Prausmüller 2011). But with clear demarcation to this, Cox was concerned with the *social forces* acting in the global sphere, which, in a Marxist tradition, would mainly entail class relations. However, he also introduced the term of *social groups* and therewith opened the concept for all collective actors pursuing the assertion of their interests or counteracting against a hegemonic constellation. Remaining close to Gramsci's original idea on the state as the basic entity in international relations, Cox developed the concept of the "internationalization of the state" (1983, p. 169). Therein, he identifies the state as the "primary focus of social struggle and the basic entity in international relations" (Cox 1983, p. 169).

Accordingly, in the very first instant, hegemony would have to be established at a national level. However, since it is the nature of the capitalist system, this dominant mode of production would then penetrate into other nation states, that is "nation-based developments which spilled over national boundaries to become internationally expansive phenomena" (Cox 1983, p. 169). Thus, Cox understands world hegemony as an "outward expansion of internal (national) hegemony established by a dominant class. The economic and social institutions, the culture, the technology associated with this national hegemony become patterns for emulation abroad" (Cox 1983, p. 171). It is a complex of international social relations which connects social classes on a global scale. In this process of expansion, hegemony is to be reproduced and the political-moral leadership maintained. It is through mechanisms of regulation and the dissemination of universal norms and values in international institutions that global social contradictions become processable and which shape the behavior of states and civil society actors across borders. At this stage of international hegemony, hegemony and its corresponding ideologies are mainly developed and reproduced at the level of international organizations (Cox 1983).

This innovative departure from classic International Relations constituted a starting place for a neo-Gramscian perspective in IPE (Opratko and Prausmüller 2011). What today became a broader field of research, initially was shaped by three main work strands of neo-Gramscian perspectives (Scherrer 2005; Overbeek 2003; Opratko and Prausmüller 2011) which Bieling (2011b) labels as follows: the critical-realistic perspective initially developed by Cox, the transnational perspective of the "Amsterdam school" mainly led by Kees van der Pijl, and the constitutionalist perspective developed by Stephen Gill. The common aspirations of these different approaches, and of neo-Gramscian work in the IPE in general, mainly lie in the intention to capture the connectivity and interrelations between structures and agency and their historic grounding. By acknowledging the dialectic between these, neo-Gramscian scholars reject a bias towards either structuralist or actor-oriented approaches (Overbeek 2003, p. 169). As Cox put it already in his early work:

"Hegemony at the international level is thus not merely an order among states. It is an order within a world economy with a dominant mode of production which penetrates into all countries and links into other subordinate modes of production. It is also a complex of international social relationships which connect the social classes of the different countries. World hegemony is describable as a social structure, an economic structure, and a political structure; and it cannot be simply one of these things but must be all three. World hegemony, furthermore, is expressed in universal norms, institutions and mechanism which lay down general rules of behaviour for states and for those forces of civil society that act across boundaries—rules which support the dominant mode of production" (Cox 1983, pp. 171–172).

Accordingly, neo-Gramscian IPE scholars approach hegemony as a consent-based mode of transnational socialization which includes class affiliations, ideological relations as well as structures of power and domination (Bieling 2011a, p. 12). Other common grounds of neo-Gramscian scholars include the research interest in understanding the deepening of capitalist relations of production, the political articulation of class interests or the evolution of transnational social relations and hegemony (Overbeek 2003, p. 172 ff.).

The Amsterdam school centered its work around the analysis of the process of transnational class formation. Within the reproduction cycle of capital, class fractions would organize themselves around their functional differentiation. This process, however, was not regarded to be merely economic, but rather also a political one. Following in the footsteps of Gramsci, the need for the capitalist Bourgeoisie to establish a universalized notion of the "general interests of capital" was seen as an important element in the analyses. This process of universalizing the interest of a particular class has been borrowed from Gramsci's concept of becoming a hegemonic power (Opratko and Prausmüller 2011, p. 22). The Amsterdam school's scope of analysis was at the transnational level. Thereby the world order with its particular historic moments of establishment of new modes of control by a particular dominant transnational class were at the center of interest. Amsterdam scholars see in the prevailing neoliberal world order a successful project of hegemony which is significantly ensured at the ideological level by determining and normalizing neoliberal agendas (Opratko and Prausmüller 2011, p. 22).

In contrast, Gill (2011), who also analyzed the establishment of the neoliberal world order from a neo-Gramscian perspective, identifies too much coercion in the neoliberal project and therefore objects to the diagnosis of a hegemonic neoliberalism. Even though today's world economy is characterized by the internationalization of production and finance and the spread of a consumerist ideology, due to its many system inherent crises, global capitalism would be relying on many measures of coercion. Indeed, global capitalism has reconfigured local social structures and forms of states while integrating national economic systems worldwide. But nevertheless, the contradictions that the global capitalist system constantly create, being structurally based on the exploitation of mankind and nature, would make the system weak and questionable. Ecological crises, the restructuring of production and increasing economic disparities leading to global mass migration are some of the crucial crises generated by the system. Gill identifies a global power constellation which would build the basis of a transnational historic bloc. The block, however, would have to constantly work towards the maintenance of the porous system and the extension of capital power. In his concept, governments from the G8/G20 group as well as networks of private power constitute the two important groups in the block. In this context, Gill has elaborated on the relation between states, markets and the power of capital (Gill 2008, p. 100 ff.). Partly building on French regime theory, he points out that "capital as a social relation depends on the power of the state to define, shape and be part of a regime of accumulation" (Gill 2008, p. 104). Nevertheless, business power would have increased over the past decades and is exerted in direct and structural dimensions. The former, direct power, includes influence and power over labor and financial resources, contacts with governments either in the form of experts and advisors or through lobbying, and control over or influence on the mass media. The latter, structural power, is linked to material and normative dimension, most importantly, market structures and ideology which might intertwine in such assumptions that the achievement of growth would be fundamentally dependent from investment and innovations by the private sector (Gill 2008, p. 105). In the Gramscian tradition, Gill sees at the core of the transnational historic block an ensemble of social, economic, cultural, and political forces that seek to lead global neoliberalism. They consist partly of state institutions and gigantic TNCs which also provide the material basis of the block. In addition, individuals who became extremely rich with financial services and hedge funds, but also midsize enterprises, subcontractors, stockbrokers, consultancy companies, etc. together form the block. Gill describes the block's structures as transnational in scope and reach and working at all scales of capitalism (Gill 2008, p. 105).

Neo-Gramscian work has been criticized from several angles, but the transnationalization of hegemony, the fixation on elites, and the neglection or underconceptualization of the nation state are among the most recurring criticisms (Scherrer 2005; Opratko and Prausmüller 2011, p. 23 ff.; Brand 2007). One of the major arguments concerns the transferability of Gramsci's core ideas of hegemony and civil society to the international realm (Germain 1998). By emphasizing the point that Gramsci would have been a "national thinker" they caution against removing his concept from the historic context and applying it to a different frame (Opratko and Prausmüller 2011, p. 25). This argument has been contested mainly by drawing attention to several passages of the Prison Notebooks in which Gramsci treats and conceptualizes international aspects of hegemony. Furthermore, neo-Gramscian work would not seek to precisely transfer but rather to apply major elements of Gramsci's body of work to IPE theory building (Opratko and Prausmüller 2011, p. 25). Another major point of critique concerns the fixation on civil society processes while systematically underestimating the role of state actors or international treaties between states (Burnham 1991, p. 88). Brand (2007, p. 163) makes the point that neo-Gramscian work often tends to privilege classes against the state. Moreover, particularly a fixation on a capitalist elite in some influential neo-Gramscian works as for instance by Gill would attribute too much importance and ability to influence to global capitalist elites. According to this strand of critique, this analytical underestimation of nations states in the process of globalization positions states as a passive victim during the increase of structural and organizational power of global capital. At the same time, potential and ongoing processes of counter-movements, resistance and contestation would remain underappreciated, too (Scherrer 1998). Today there is a broad strand of neo-Gramscian work that puts a particular emphasis on the further development of the concept on the "internationalization of the state". In Buckel and Fischer-Lescano's (2007) reader *Hegemony armored with coercion* a whole set of approaches integrating state theory into a neo-Gramscian concept is published. Therein, Ulrich Brand elaborates on the internationalization of the state. Just as in Gramsci's writings, the integral state bears the function of stabilizing the societal status quo, at the international level, international state apparatuses and organizations would fulfil this very role. With what Brand calls the compression of second order, he describes a process in which particular interests are transferred to and established at the international level. Accordingly, firstly, at the national level, particular interests consolidate to state policies for which will be, secondly, advocated for at the international level. A clash with the interests of other states and, importantly particular interests of civil society groups, will also take place in this process at the international level. In this process of contention, the compression of second order will achieve an instable but somehow hegemonic constellation which represents the contemporary balance of power and is institutionalized in the major international state apparatuses and organizations. The special role of civil society, interpreted as a contested space of struggle over influence and consent, in this process is recognized here, too. The establishment of generalizing production and consumption norms would mainly take place at this level. However, one of Brand's main points is that in most cases, non-public actors representing private interests would also have to align and articulate those interests with the prevailing modes of intergovernmental political practices. Nevertheless, he also acknowledges the possibilities of private actors to pro-actively exert influence on them.

Apart from the critique on transferability to the global sphere, critiques have been brought up to challenge the underlying theoretical assumptions. While from a Marxist-orthodox perspective neo-Gramscian work is criticized for not being able to grasp the essence of the global capitalist socialization, from a post-structuralist perspective neo-Gramscian work's emphasis on economic relations is seen as problematic (Scherrer 1998; 2005). In their influential writing on hegemony and radical democracy Ernesto Laclau and Chantal Mouffe (1991/2020) confront Gramsci's theory with the reproach of not overcoming Marxist "essentialist apriorisms", such as economism, etatism, and classism (cf. Scherrer 2005). The privileged position of the economic base would lead to an underappraisal of cultural, discursive, and ideological mechanisms of power establishment. Accordingly, economism would therefore imply that economic development is regarded to show distinct political effects where classes are set *a priori*. The notion classism draws attention to the reproach that collective identities and interests in the capitalist society would be determined through the respective position in production (Scherrer 2005). Where some neo-Gramscian authors reply to such critique by integrating post-structuralist elements and laying distinct emphasis on the specific connection between the hegemonic dimensions of the political, ideology, law and culture (Wullweber 2014; Worth 2008), others contest the assumption that the focus on economic classes would automatically imply an economist reduction of societal complexity (Cox 1983; Morton 2006; Opratko and Prausmüller 2011). Morton (2006, p. 66) highlights that it would even be a particular feature of neo-Gramscian work to reflectively approach the notion of class and to recognize its multicausal ideological and political moments.

Besides such theoretical claims, other critiques concern more empirical aspects such as scale and agency within neo-Gramscian work, which are seen to be underdeveloped (Brand 2007; Winter 2011). Similar to other theoretical approaches in International Relations, the neo-Gramscian work recognizes the existence and importance of geographical scales but analyses them as predefined spheres which were not also object of change and negotiation. Brand (2007, p. 163 f.) argues for an understanding of spatial scales which would be constituted through processes of social interaction, similar to approaches developed in the field of Critical Geography. Accordingly, regimes of governance would be divided into different dimensions (local, national, regional) according to socially agreed upon criteria. In many empirical studies, he argues, the interconnection between these scales does not receive much attention. The present study seeks to respond to this strand of critique by focusing on the interconnection between different production segments and their respective governance and institutional environment as is practiced within the scope of GVCA. Winter (2011), equally highlighting the importance to empirically grasp the different scales of a given hegemonic constellation, argues for an improved analysis of the relationships between the various scales and actors, an argument that has been supported by Scherrer (2005) who has criticized neo-Gramscian theory as generally being weak on the point of defining societal actors and the scales in which they act.

One approach that seeks to more precisely define the actors within the "black boxes of social contestation" (Winter 2011) has been developed by Leslie Sklair. In order to characterize what he calls "the transnational capitalist class", Sklair highlights three main dynamics which would be constituent to it: 1) TNCs became the main institutional form of transnational economic practices, 2) a transnational capitalist class is evolving in the global political sphere, and 3) the culture-ideology of consumerism became the prevailing one (Sklair 1997, p. 520). Applying a Marxist language of class formation, he goes on to define four groups within the *transnational capitalist class*. Accordingly, there is (1) a corporate fraction which owns and controls "the major transnational corporations and their local affiliates" (Sklair 2003, p. 3), (2) the political fraction of globalizing politicians and bureaucrats, (3) the technical fraction consisting of globalizing professionals, and (4) a consumerist fraction comprising merchants and media (Sklair 2003). Moreover, Sklair describes close relations between these fractions and civil society institutions. In many cases, the personnel in these groups would be interchangeable. Particularly senior members would often belong to several fractions at the same time and also hold key positions in boards of Think Tanks or NGOs (Sklair 1997, p. 521 f.). Another key feature of these fractions would lie in their complementary functionality. Organized in a transnational governance network, they would be simultaneously working at different geographical scopes towards the common goal to maintain and expand structures, institutions and ideologies that favor the neoliberal project.[2] Having a closer look at Sklair's depiction of actors involved in the leadership of "the transnational capitalist class" reveals a strong congruency with the societal forces that Gill identifies to be constituent to the transnational historic block- that is economic, political, social, and cultural forces. For the present study which is primarily concerned with the perspective of TNCs and their activities to maintain and extend control and influence, particularly through the implementation of CSR strategies, it is important to include a clear picture. It is therefore important to define what is actually meant when talking about a transnational history block and which actors are actually meant when talking about societal forces. Although the present study does not apply a class determinism as outlined by Sklair with his theory of the

---

[2] More insights in the network character of global capitalism has been elaborated by the French Sociologists of the "Pragmatic School" as eg. Boltanski and Chiapello, cf. Demirovic 2008.

transnational capitalist class, it follows the terminology of a transnational historic block, recognizing the overlaps between the two approaches, and acknowledges Sklair's concrete spelling-out of typical actor groups as an important element to appropriately grasp the "black boxes of societal forces" in the transnational historic block.

What are the main insights gained by the presented confrontation of the different approaches within the neo-Gramscian strand in the field of International Political Economy and the critiques on them? What mainly becomes clear, neo-Gramscian work, all following the same objective to describe and explain processes of power reproduction within a global scope, takes two different paths and is important to position the present study in it. While the strand of the "internationalization of the state" recognizes the existence and importance of private and civil society actors, the main locus of struggle over influence and domination is allocated to the international state apparatuses and organizations (so as Cox and Brand). The starting point for the internationalization of hegemony is the outward expansion of strong national hegemonies. The internal configuration of international institutions and the rules and regulations they develop represent the power balances in the international power balance. In this process, private actors also have a stake but have to adapt to the governmental modes of the international community. In contrast, neo-Gramscians focusing on a transnational historic block (as eg. Gill, Sklair) go one step beyond and exert a real break with state-centrism. National states become only one actor group among many and the focus lies more on the ensemble of power players in the field of global governance. The present study recognizes the importance of both approaches and assumes that reality might lie somewhere in the middle of both strands. While international organizations still play the main role in international politics, many network-like, primarily private institutions gained importance over the past decades and important agenda are adopted at these levels which have distinct influence on e.g. UN, EU and national policy processes.[3] Hence, the "internationalization approach" seems to somehow underestimate the diverse global governance forums and the direct and structural power dimensions of global business actors. At the same time, the "transnationalization strand" lacks an adequate and systematic integration of public and private spheres at any of the given governmental scales. The present study seeks to respond to this deficit by paying attention to relations between private and public actors in the Ghanaian cocoa sector which evolve in the course of the implementation of CSR projects.

---

[3] For example, for a comprehensive empirical study on the global governance constellation of water see Dobner 2016.

Critiques on economism and class reductionism seem appropriate but have been replied to by neo-Gramscian scholars quite convincingly. A careful and empirically informed conception of the transnational historic block seems to be an opposite answer to overcome such ideology-based controversies. It puts a distinct emphasis on societal groups and forces involved in global governance practices. The use of scales and agency are a good entry-point for such an undertaking and therefore has been integrated in the design of the empirical study (see chapter 6). Finally, what the presented theoretical streams have in common and what is the starting point for the transmission of neo-Gramscian concepts of hegemony to the study of CSR in global value chains, is the perception of the neoliberal world order as being in a fundamental, but capitalism-inherent, continuous crisis. The porosity of hegemony at the global level (Candeias 2008) would lead the block to constantly struggle to reproduce itself (Sklair 1997) by applying measures of consent and coercion, which, in turn, have been theorized in different ways again. Demirovic (2008), for instance, identifies too many elements of coercion applied in order to keep the neoliberal world order so that he denies it a status of hegemony at all. Similarly, Gill (2011) has conceptualized the block's strategy to sustain global capitalism despite its many contradictions and crises with a notion he has called "new constitutionalism". Aiming to enshrine the primacy of a neoliberal world order, "new constitutionalism" comprises three main partly coercive measures: 1) Measures to re-compose the states apparatuses in a way that they follow the logic and needs of transnational capital, which he illustrates with the creation of independent central banks in which capital interests would be strongly represented in the executive boards, 2) measures to create and extend capitalist markets, e.g. through the commodification and privatization of natural resources, and 3) measures seeking to handle contradictions in the system and arising counter-movements to it (Gill 2011, p. 269 f.). Gramsci's concept of passive revolution and therein the approach of *trasformismo* become central here. With *trasformismo,* measures to keep legitimation of the system are applied equally as measures of compensation and the co-optation of political opposition (Gill 2011, p. 270). All three are central functions that CSR, as both an economic and a discursive strategy, fulfils in the struggle to maintain and increase control and influence in a given supply chain and its institutional environments. The notion of *trasformismo* is therefore among the most important elements of Gramsci's work applied in the present study.

3.2.3   CSR, Trasformismo, and the Manufacture of Consent in Global Value Chains

While the above-outlined studies seek to analyze global power formations and the functioning of the hegemonic project as such, the present study puts emphasis on one particular element within these societal struggles: CSR as a consent strategy. As a consent strategy, CSR comprises several dimensions which have to be recognized, but not all will receive detailed attention here. Based on the theoretical insights gained during our discussion of neo-Gramscian perspectives in IPE, an understanding of societal dynamics of domination, the role of transnational social forces which benefit from the neoliberal global order, the meaning of instability, consent creation and other elements of the struggle to keep and foster the dominant position has been laid. Recognizing the central position of global value chains in the contemporary global economic system and the driving role TNCs play in them, this subchapter carves out the relationship between CSR, global value chains and the manufacture of hegemonic consent as the central background for the subsequent empirical study of CSR as a governmental tool in the GCCC. For this discussion, the question whether the current neoliberal system is to be understood as a hegemonic formation or a situation of supremacy is secondary and will not be discussed further. Nor will the question of whether the current formation of forces that has been theoretically described as a transnational historic block should be considered as a transnational capitalist class or not gain further attention here. These questions do not immediately concern the major research interest of the present study. As explained in detail above, the interest lies in the description of CSR as a tool for maintaining and expanding power positions and in the understanding of its transformative potential in a given chain, particularly at the local level of production. For this purpose, Gramsci's notion of *trasformismo* is particularly valuable and paves the way for an interpretation of CSR as a consent strategy. Horn (2011) gets to the heart of the notion by summarizing CSR as "a strategy of hegemonic maintenance of power through political integration of subaltern groups and through concessions and compromises of the leading social groups" (Horn 2011, p. 210). She particularly highlights the often rather neglected role of subaltern groups in this process. Measures of *trasformismo* as a consent strategy within the larger project of a passive revolution (see section 3.2.1.) only become necessary because subaltern groups continuously challenge the hegemonic project which therefore has to be renegotiated all the time (Horn 2011).

Often based on such a dialectical understanding of consent, several studies have approached CSR as a form of "double movement" (Utting 2005; Levy 2008;

Levy et al. 2010). In this context, CSR is recognized to be initially used by various subaltern groups, particularly civil society groups such as trade unions, NGOs or activist groups, seeking to criticize social and ecological irresponsibility and to claim for binding business obligations (see for this discussion chapter 2). Curbach (2009) has shown how business actors, during the process of global neoliberal agenda-setting, have managed to capture the formerly critical potential of CSR, to integrate it into their own agenda, and to neutralize it to a large extent. Horn's work (2011) gives insight in how CSR lost big parts of its critical potential at the level of the European Union and served as a measure of *trasformismo* there. Horn, equally as De Schutter (2014), highlights the point that the European Commission (2001) came out with its green paper *Promoting a European framework for Corporate Social Responsibility* as a complement to the Lisbon process during which the competitiveness of the European economy was sought to be increased by measures of liberalization and privatization. She describes the subsequent setting-up of a European multi-stakeholder forum on CSR as an important step of *trasformismo* through CSR. While the forum's mandate did not include the elaboration of concrete advice, from the outset, the commission supported the position of a voluntary approach to CSR and important claims from civil society groups remained unreflecting (De Schutter 2008/2014, p. 213, cited after Horn 2011, p. 214). This, according to Horn, would have created a pseudo-participative process. While the establishment of a multi-stakeholder forum suggests a deliberative and participative political process, claims opposing the project of increasing market liberalization and related reforms do not receive significant attention but are all brought under the umbrella of CSR. By creating the forum, subaltern groups are integrated in the project. The confirmation of the need for CSR by the commission can be regarded as a partial concession to counter-movement claims while incorporating and absorbing the fundamental critiques in the own discourses (Horn 2011, p. 214).

   This sketch of CSR's effects as a means of *trasformismo* at the level of the EU allows to gain a first idea on its transformative potentials in other contexts, too. Following the goal to achieve a broader understanding on how CSR can be assessed as a hegemonic tool which is employed to foster and achieve consent by lead firms in a value chain, it is important to reflect on the different scales it is debated at and finally put into practice. While much attention is given to its consent-creating function at higher transnational and regional governmental levels, its consent-creating function at the local level of implementation has gained almost no attention so far. Whenever focusing at the local levels, analyses of CSR' implementation mainly investigate the immediate outcomes such as benefits for the targeted groups or at the company level. Nevertheless, the three *trasformismo* dimensions carved out by Gill (2011) presented above (measures to keep legitimation of the system, measures of compensation, and measures of co-optation of political opposition) have to work simultaneously at all different scales of the hegemonic project. Furthermore, a neo-Gramscian perspective also suggests that, in order to pursue the strategy of consent creation next to the economic and political dimensions which gain most attention in the presented studies, also social, institutional, and cultural dimensions have to be considered when empirically assessing its modus operandi.

Next to the different scales and dimensions of the hegemonic project, actors need to be assessed distinctly. Confronted with the several black boxes of societal forces (Winter 2011: 153), not all of them can be opened and operationalized in one single study. As described in several places in the present study, the main objective here is to explain CSR as a governmental tool for TNCs. In the empirical part of the study, relations and interlinkages between TNCs and other sector stakeholders at the local in Ghana will be shown and discussed. However, it is not the topic of the study to describe the whole formation of the transnational historic block and to describe dynamics at all different scales of action. Approaching the discussion from the neo-Gramscian inspired global governance perspective, the empirical part of the study exclusively addresses the local level of CSR implementation. Thereby, it stands out from most of the CSR analyses in so far as it goes one step further and argues that CSR is also a tool for TNCs to engage in compromises with subaltern value chain stakeholders at the local level and to align/transform local structures and institutions in a way that is facilitating the process of keeping and gaining control within the respective chain. The innovation of the present study lies in the combination of these neo-Gramscian inspired governance features with the industrial governance understanding of Global Value Chain theory. The nexus of global and industrial governance analyses allows for the integration of the crucial dimension "control over production flows" in the concept of CSR as a consent strategy. The following subchapter sets out the combination of two concepts and their application to the purpose of the study.

3.3   Governing through CSR—Elements of the Consent and Control Framework

The assumed high potential of CSR to serve as a governance tool for TNCs stems from its above-described umbrella function. Most of the stakeholders in the GCCC agree with the claim that TNCs should assume more responsibility in the chain. The voluntary support to solve sustainability issues in the chain and the assistance of farmers to boost their production capacities is highly welcomed by actors from diverse stakeholder groups. The above-presented hypotheses give rise to the concern that CSR actually has an even deeper effect than one might expect at a first glance. One of the central arguments of the present study is that CSR leads to profound societal and institutional transformations, mainly, though not exclusively, in the target areas. If these claims can be reaffirmed with the empirical study, the mechanisms of the transformative processes would need to gain further attention since lasting shifts in power constellations and local structural change will likely result. In order to provide a solid tool for such a critical analysis, in this subchapter CSR as a governance tool is conceptualized. Its diverse soft and hard tools of governance will be framed as tools for consent and control.

As presented in section 3.1, the GVCA approach is a strong instrument to conduct a systematic analysis on a given value chain and to develop a detailed description of the main aspects of its functioning. It helps the researcher to unfold the organization of industry procedures and dig into its characteristics, like the respective endowments and capacities of the stakeholders. Main questions that guide the analysis are concerned with the location of important production steps, how they are coordinated between the different production segments, where the main share of profit is going, and what can be done to integrate producers and boost their position in the chain (keyword upgrading). When studying global value chain governance, particular emphasis is given to the tools and strategies lead firms apply in order to coordinate and control the resource flow between the segments. Typical instruments for these governmental activities are process and product standards, contracts, and the set of different relationship qualities with suppliers which has been described in section 3.1. Some of these elements of value chain governance are also important measures in the frame of CSR strategies. Particularly social and environmental process standards overlap with CSR contents. The empirical part of the study will show that not only these, but also contracts and relational shifts are important aspects of CSR. Therefore, the conceptual framework builds upon the analytical strength of GVCA to carve out how TNCs exercise control over the production flow in their chains. Nevertheless, GVCA is limited in its ability to describe CSR as a governance tool. Most importantly, its exclusive focus on only industrial aspects of governance leaves a huge explanatory gap. The approach comprises an institutional axis. But, it is presented and applied as something external from the chain which impacts on its functioning but is not in reach of lead firms' activities. Hence, activities of TNCs which seek to shape the institutional environment of their chains are completely excluded from the analysis. Yet, the particular potential of CSR as a governance tool is to respond to the double pressure by also managing societal pressures and establishing enabling environments for their value chains.

This is why next to an understanding of governance as the control over resource flows the conceptual framework equally builds upon the idea of governance as a consent strategy. The neo-Gramscian perspective on global governance and how it helps to conceptualize CSR as a tool to create consent and an enabling environment for value chains has been developed in the previous subchapter. The strength of this perspective is that it allows researchers to open up their view on governance activities and become sensitive for soft tools including efforts of dominant players to influence institutions and perceptions and attitudes of stakeholders. Moreover, this approach provides the researcher with a number of tools to identify how TNCs might try to respond to crises and pressures which jeopardize the status quo. In section 3.2 governance activities have been outlined as measures of *trasformismo* and include activities which seek to stabilize the system or restore the legitimacy of an actor. Prominent instruments for this type of governance include the establishment of new forms of dialogue which serve to disseminate and universalize particular interests and meanings and values. Many global value chains are undergoing multiple social and environmental crises and are facing a variety of risks and pressures. Some dimensions of CSR respond directly to these challenges and may function as such measures of *trasformismo,* aiming to stabilize the system surrounding the given chain. In this context, CSR as a consent strategy could comprise, for instance, of activities to work towards compromises with subaltern members in the chain. Thereby, as outlined above, CSR has the potential for co-opting critiques which arise on the current status quo of crises-causing situations such as severe poverty and ecological degradation, articulated by civil society groups or local dwellers. Finally, in the course of their participation in CSR projects, subordinate groups such as producers and civil society actors might be partially compensated and some material concessions might be given. In the course of such a process of consent creation, subalterns might develop a real interest in the CSR project and expect real benefits from their participation in it (cf. for elaborations on such dynamics Candeias, p. 20). However, in order to implement CSR strategies successfully, an enabling institutional environment and the political support at all concerned levels are required. TNCs therefore have to engage at these levels, too. They have to build alliances with key players and create the supporting structure for the successful implementation of the CSR agenda.

Winter (2011) stresses the importance of both the role of particular actors and levels of action. The main guiding questions that he suggests ask 1) how a hegemonic aspiring (or fostering) societal group manages to generalize its particular interests and 2) which social actions they thereby apply. In order to appropriately grasp the process of establishing or fostering the hegemonic project, the spatial-functional differentiation of these processes become important: At which scales do stakeholders act and in which relation stand these scales to each other? Which actors become active on which particular scales? And besides, which are the topics and objects of conflict and mediation at the various scales? It becomes visible here how GVCA can help to systematically reply to these questions. The strength that the focus on governance through consent creation adds to the concept is the inclusion of the following main levels of analysis that Winter proposes: the actors of the process including non-industrial actors, the scales of actions which are not necessarily the same as production segments, the means of action including non-industrial activities and the object of struggle/topic of active consent creation in a societal context. Figure 3.2 presents the conceptual framework on governance as consent and control strategies which is based on the integration of the two complementary research strands.

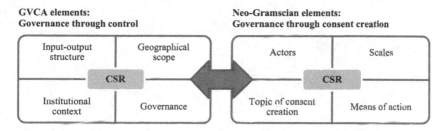

Figure 3.2 Axes of the consent and control framework. (Source: own elaboration)

In the following, this conceptual framework will be applied to the global, the national, and the local levels of cocoa and chocolate production. At each level, an overview of the composition of chain functioning and main actors is provided as well as CSR activities located. The eight axes will not gain attention at each level, only at those which seem relevant for the study of CSR at the respective level. It is in the empirical part of the study where the framework will be applied in detail. CSR governance activities which are relevant at the local level and which will be assessed based on the framework, among others, comprise production control measures such as standards, input delivery schemes and the linked creation of new institutions. It will be considered, for instance, which areas are mostly targeted by CSR and how the perceptions regarding quality and quantity of the production alter with their participation in a CSR project. Typical consent

creation components such as the shaping of the sustainability discourse or concessions which allow for livelihood improvements will be important aspects of
the analysis at this level. Thereby, it will become clear that the analytical categories outset here are often overlapping and many measures are directed towards
both sides of the double pressure to which CSR seeks to respond. By undertaking this analysis, the framework is only meant to guide the research process and
is not to be falsified by empirical evidence. The justification of such a research
strategy is given in detail in chapter 6 on the field work design. Before we turn
to this, an overview on the composition and functioning of the GCCC and the
cocoa industry in Ghana, the main challenges, and the different forms of CSR
applied to them (chapters 4 and 5) shall lay the basis for the understanding of
challenges at the local level of cocoa production.

CSR in a Dynamic Global Cocoa-Chocolate Chain

4

For the investigation of CSR's potential to serve as a tool for improved control and consent in a given supply chain, the GCCC is an instructive and fruitful example for several reasons. As already highlighted in the introductory chapter, the final good, that is chocolate and other confectionery, has a high emotional value attributed to it. The contradiction of consuming a product that has such a high positive emotional value and while possessing the knowledge of poor working conditions and severe poverty among cocoa producers, causes concern for many consumers. This often translates into the willingness to pay a price premium for Fair Trade (cf. Vlaeminck et al. 2016) other sustainability certified products which increasingly are results of transnational CSR strategies. Furthermore, cocoa production being rooted in the colonial past, the traditional divide of cocoa production based on the shoulders of millions of poor peasants in the Global South and cocoa consumption initially reserved to the affluent and nowadays by the masses in the Global North, brings the global justice perspective even more to the forefront. Charlery de la Masselière (2014, p. 129) has drawn attention to the role of the introduction of cash crops in rural Africa and how they served to legitimize and institutionalize hierarchies of division of labor downstream the supply chains from the Global South to the Global North. In the GCCC, similarly to most other agricultural supply chains too, this asymmetric labor divide is mirrored in a significant asymmetric value distribution along the chain, what is another matter of injustice. These and other sustainability issues oblige the transnational cocoa-chocolate industry to put more efforts in improving local conditions of production because the industry is simply depending from a functional cocoa production. Hence, CSR is strongly practiced and plays an increasing role in the GCCC. The objective of this chapter is to understand the composition and workings of the GCCC, reveal major governance trends, show

F. Ollendorf, *The Transformative Potential of Corporate Social Responsibility in the Global Cocoa-Chocolate Chain*, (Re-)konstruktionen – Internationale und Globale Studien, https://doi.org/10.1007/978-3-658-43668-1_4

which CSR activities are conducted and how they link to the major challenges in the chain. The analytical framework developed in the previous chapter guides the analysis in this chapter. It provides detailed information on the functioning of the GCCC, discusses important shifts in its institutional environment and presents the major challenges for cocoa production. The interplay of control and consent measures becomes most visible in subchapter 4.3 on the hybridization of governance in the GCCC. These insights provide the background for a good understanding of the function of CSR in the chain.

The chapter proceeds as follows: after a brief introduction to the main features of the GCCC such as history, geographical scope, and input-output structures of the GCCC, main production segments and patterns of governance will be presented. Subsequently, the chapter turns to the most pressing sustainability challenges in the GCCC. For each of the three sustainability pillars, environmental, economic and social sustainability, attention will be given to one core problem of each pillar. Finally, the chapter closes by giving an overview of the different approaches and forms of CSR in the GCCC and how they relate to the presented challenges. This provides the basis for the following chapter on CSR in Ghana's cocoa industry will.

4.1 Historical and Geographic Features of the Global Cocoa-Chocolate Chain

The consumption of processed cocoa products dates back to the Olmecs, Mayans, and Aztecs who used cocoa for a hot and spicy drink during religious celebrations. Yet, cocoa originates from the Amazonian lands and arrived to Mesoamerica via merchant relations. After the arrival of the Spanish colonizers, cocoa was brought from Mesoamerica to Europe in the 16th century. It remained an exclusive drink consumed only by aristocracy and clerics for the next two centuries. Mainly because of its elaborate manufacturing process which required rare spices such as cinnamon and vanilla to be added, cocoa took until the 19th century to spread over to other parts of the European populations—much later than tea and coffee (Durry and Schiffer 2012, pp. 11–12). The increasing demand made chocolate production a lucrative business which led the colonizers and traders to push cocoa production in the colonial territories. First, the colonial cocoa production was expanded to other parts of Latin America like regions such as present Ecuador, Venezuela, and Brazil where huge plantations were installed. As the indigenous people were forced to work on sugar, tobacco, and cocoa plantations under extremely perilous conditions which decimated their populations drastically, slaves were brought from West Africa to work on these cash crop farms. Hence, this triangular trade system became the major source of the colonial powers' wealth accumulation. Cocoa spread to Asian colonies by the Dutch and Spaniards in the late 17th century and was brought to Africa only in the 19th century. The first cocoa production in Africa was initiated by the Portuguese at Sao Tomé and Principe in 1820 (Poelmans and Swinnen 2016, p. 17). At the end of the 19th century cocoa arrived to Côte d'Ivoire and Ghana, today's two main cocoa-producing countries. Within fifty years the production rose tremendously in these two countries and many rural dwellers began to grow cocoa.

In the late 19th century, Swiss and Dutch technological advancements enabled for product diversification and new products such as milk chocolate emerged (Clarence-Smith 2016, p. 58; Fromm 2016, pp. 74–75). During the same period, urbanization, decreasing transport costs, and a fast augmenting purchasing power in European countries triggered a "great chocolate boom" in the late 19th to early 20th centuries (Clarence-Smith 2016, p. 58) and chocolate entered the mass markets in Western Europe and North America. Swiss, Dutch, Belgian, German, and British chocolate manufacturers began to invest in product innovations and commenced production of new types of chocolate products like chocolate bars or pieces. Cocoa for mass consumption was now imported from West Africa, mainly from the Gold Coast (current Ghana) which supplied 40% of the raw material needed to meet the increasing demand for chocolate (Muojama 2016, p. 5). Whereas in Latin America cocoa was initially grown on huge plantations, cocoa in West Africa was produced by small-scale farmers from the beginning (Durry and Schiffer 2012, p. 39). During the first part of the 20th century, cocoa became one of the major income sources of the producing colonies and a central occupation for thousands of rural dwellers. However, cocoa world market prices were volatile and often very low. The livelihoods of most small-scale cocoa farmers remained very poor.

After West African countries regained their independence, the general patterns of the industry remained in place, with the exception of the governance structure. The picture of a colonial crop, where underpaid small-scale farmers and plantation workers from the Global South work hard for the production of a luxury good mainly consumed in the industrialized world and lately also in the transition economies till today largely holds true. For instance, in Côte d'Ivoire and Ghana, cocoa production remained a backbone of the economies and received substantial state support. Only with the introduction of the Structural Adjustment Programs (SAPs) by the World Bank and the IMF in the 1980s and 1990s did some governments of the main West African cocoa producing countries remove state support and liberalized the sector. These policy shifts did not lead to a general improvement of cocoa farmers' situation.

Since the introduction and rapid growth of cocoa up until today, Côte d'Ivoire and Ghana have kept their leading positions in the world cocoa market. In the cocoa production season 2018/19, about 70% of the whole world cocoa production came from only four African countries: Côte d'Ivoire, Ghana, Cameroon and Nigeria. In 2015/2016, the two countries together accounted for 60%. 17% came from Latin America and 10% from Asia (International Cocoa Organization 2017). In 2018/2019, the world's top five cocoa producers were, as visualized in figure 4.1, Côte d'Ivoire (2,180,000 metric tons in 2018/19), Ghana (812,000 metric tons), Ecuador (322,000 metric tons), Cameroon (280,000 metric tons), and Nigeria (250,000 metric tons) (Statista 2020a). Over the past years, Côte d'Ivoire sharply increased its production which in 2016/2017 was 600,000 tons higher than three years before (Fountain and Hütz-Adams 2018).

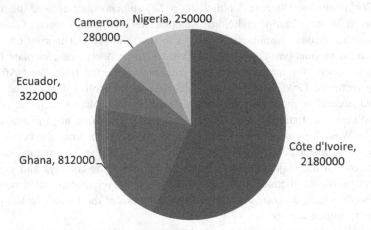

Figure 4.1 The top five producing countries' production in metric tons (2018/2019). (Source: Statista 2020b)

Over the past decade, both cocoa production and consumption steadily increased. Importantly, the retail consumption of chocolate confectionery worldwide was continuously growing from 6,946 thousand metric tons in 2012/13 to an estimated 7,696 thousand metric tons in 2018/19 (Statista 2020b).

Traditionally, Europe and the United States are the main consumers of chocolate confectionery products. However, this picture is changing as the main emerging economies have increased their chocolate confectionery consumption. By 2020, Russia is estimated to contribute 11% to the global consumption just after the United States which is estimated to lead global consumption with 18%. Brazil (4%), India (3%), and China (3%) are estimated to also rank under the top ten of chocolate confectionery consuming countries directly after the United Kingdom with 8% and Germany with 7% of the whole world consumption (Statista 2017). But also, there is a new demand for chocolate products within economically advanced African countries. Nigeria ranks third in terms of the world's fastest growing chocolate markets and realized a consumption growth of 286% from 2007 to 2012. South Africa ranks among the eight countries which account for 70% of the world's confectionery growth (KPMG 2014, p. 3). New regions of consumption bring along new types of demands and requirements facing the industry. For instance, Mondelez International patented a chocolate that does not melt for three hours in a 40°C environment—a ground-breaking innovation for consumer markets with hot climate conditions. Simultaneously, consumers from traditional markets have diversified their preferences. There are trends to increased consumption of luxury chocolate and confectioneries with fine flavors using artisanal ingredients. The industry trend goes towards full customization from bean to bar where personalized products and consumers' knowledge on the origins of ingredients become standard. The more informed consumers tend to care more about sustainability issues in the chain so that sustainability certification and traceability systems have rapidly grown since 2010 (cf. Potts et al. 2014).

To get to any of the end-products, cocoa is the basic raw material of chocolate. The bean travels along a complex chain of production, processing, and value addition with many stakeholders having a stake embedded in the global institutional environment. Nevertheless, independently from which of the three main producing regions (Africa, Asia, and Latin America) the cocoa bean originates or whether it is an organic, sustainability certified, or conventional product, it generally goes through the same main stages of the value chain.

4.2   Main Activities within the Production Segments

As laid out in the methodical chapter on Global Value Chain Analysis, there are several ways of describing and assessing the composition of a global value chain. For the GCCC, the different stages of the production flow mainly serve as core categories (c.f. UNCTAD 2008). For instance, the World Bank (2013) subdivides the GCCC into four segments: cocoa beans, semi-finished cocoa products, couverture or industrial chocolate, and finished chocolate confectionery products. A study commissioned by UNCTAD (Gayi and Tsowou 2016) uses the main stages of activity along the chain: production, marketing, processing, manufacturing, and retailing to consumers. The World Cocoa Foundation (2014) breaks the chain down into nine activity segments: cocoa growing, harvesting, fermenting and drying, marketing, packing and transporting, roasting and grinding, pressing, chocolate making, and consumption. Any of these approaches allows for a comprehensive classification of the main activities of value-addition along the chain. The more detailed the approach, the better it is for the assessment of new activities linked to the introduction of CSR strategies. The chain is even still more complex and it is important to carefully look at each segment to see the diversity of activities undertaken there. For instance, further activities such as input supply to farmers or other marketing activities like purchasing, recording, and quality control are crucial for the sustainable flow of the cocoa beans and the functioning of the chain. As will be shown, particularly these activities are of increasing interest to the TNCs which strive for control over production sustainability and their CSR strategies increasingly shape these segments. Therefore, the detailed look at these segments is crucial for the present study because it allows for the identification of new fields of influence which open up for TNCs due to the implementation of their CSR strategies. Additionally, by analyzing these segments, it is possible to identify the resulting new responsibilities and power relations within the chain, i.e. the transformation of the chain. The local level of cocoa production and the surrounding activities are at the center of interest since these are areas which formally escaped most from the management and control of the chain by TNCs. From chapter 5 onwards, the emphasis will be on the local level of cocoa production in Ghana while chapter 4 illuminates composition, typical patterns and challenges within the entire GCCC. The figure 4.2 provides an overview of the major segments of the GCCC which subsequently will be described in more detail.

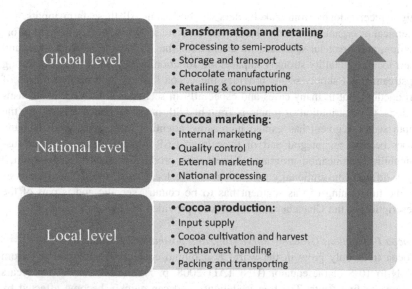

Figure 4.2 The journey of a cocoa bean: main production segments within the GCCC. (Source: own elaboration)

4.2.1   Local Level: Cocoa Production and Linked Activities

Input supply
Independently of whether they produce conventional, organic, or in the frame of a sustainability certification program, cocoa farmers will require some kind of input factors for their farm. For instance, fertilizer or plant protection products which they have to obtain from elsewhere. The availability of these products differs in the local settings and depends on the respective institutional environments in each producing country. In many cases, the purchase of input factors is a major challenge for farmers and many fake products are on the markets, especially in more remote areas. In such cases, having little knowledge of the products circulating puts the farmers at risk of using these false and possibly dangerous products. However, an increasing number of farmers are targeted by either public or private agricultural training initiatives and extension services which comprise information on input factors which enables farmers to select approved products. Another important input is the access to finance which is a challenge for many cocoa farmers. If available, particularly members of producers' cooperatives may benefit from credit facility programs of a bank or a micro credit institution or farmers

might receive loans from stakeholders of the chain. All these farm inputs, the chemical or organic fertilizers plant protection and, agricultural training and information, and financial schemes, are crucial for the sustainable resource flow and their local availability strongly impacts the functioning the whole chain. A strong upstream input supply sector provides the cocoa farmers with needed means of production. But in many cases and especially in scenarios with weak local institutions, the distribution of inputs is fraught with many problems. Therefore, the input sector segment has gained increasing attention from TNCs and in many cases become an integral part of their local CSR strategy in the form of sustainability certification programs. Since each producing country is determined by a different institutional environment at the local level, the detailed analysis of the functioning of this segment has to be country specific and is part of the description of the Ghanaian cocoa industry provided in chapter 5.

Cocoa Production: Growing, harvesting, and postharvest handling
Cocoa trees only grow in the warm and humid equatorial belt, spanning from 10°N to 10°S of the equator (UNCTAD 2008, p. 7), and need about five years before the first fruits. The tree is delicate and can quickly become affected by diseases, insects, and fungicides. Three major types of cocoa exist: the indigenous sort criollo which is more fragile and has lower yields, the more robust and more productive *forastero*, which accounts for about 80% of the world's production and hence is the main resource for the mass chocolate market, and *trinitario* cocoa, which is a hybrid of criollo and *forastero* and combines the fine flavor of the original criollo cocoa with the high productivity and resistance of *forastero*. *Trinitario* accounts for about 10 to 15% of the world's cocoa production. Hence, the world cocoa market knows two market segments: common grade cocoa for the mass market which are *forastero* beans and *trinitario* beans with *forastero* attributes and fine flavor cocoa from criollo and *trinitario-criollo* beans (Durry and Schiffer 2012, p. 38).

Cocoa is generally cultivated in two main different systems: either on huge plantations in monocultures like to some extent in Malaysia, Indonesia or Brazil and Ecuador or on small plots by small-scale farmers (Durry and Schiffer 2012, 41 ff.). However, over 80% of cocoa is grown on small-scale farms. In West Africa, cocoa is grown almost exclusively on small-scale farms with an average size of 2 to 4 hectares (Gayi and Tsowou 2016, p. 10). Also, in all other cocoa growing regions of the world, small-scale cocoa farming plays an important role. Traditionally, due to the abundance of arable land that was historically available, farmers tended to apply rather extensive than intensive farming practices. When their farm got old and the fertility of their farm soil decreased, farmers often migrated to forest areas where they could relatively easily obtain new land, settle there and begin a new farm (c.f. Knudsen and Agergaard 2015)[1]. But due to increasing land scarcity and arising land conflicts as well as agricultural training programs, over the past twenty years, a shift to intensive farming practices took place. Extensive farming, because of turning more and more virgin forest into farm plots, is often seen as the main cause of rainforest degradation. However, the intensification of cocoa farming, for instance in the high forest zones in Ghana and Côte d'Ivoire, was environmentally problematic as well. Before intensifying their farming, family farms also used their land to plant food and other cash crops; thereby maintaining indigenous parts of the eco-system. But nowadays, in many regions, particularly in high-production areas where intensification measures are applied, after a few years of cultivation, cocoa trees cover the entire plot so that agroforestry systems or mixed-culture farming are no longer possible. With the intensification trend in West African cocoa production, most of the remaining shade trees on cocoa farms disappeared (cf. Ruf 2011). Therefore, in many small-scale farming systems today, the loss of biodiversity is very high and satellite pictures look similar to the ones of plantations.

Furthermore, the shift to intensification of cocoa farming was accompanied by the introduction of modern production means such as hybrid varieties. Today, cocoa farmers mainly plant seedlings raised in nurseries and no longer use the beans of the former seasons (Amegashie-Duvon 2014, p. 17). After clearing the land and planting the seedlings, other small trees such as bananas, plantains, or palm trees might be still planted next to the cocoa seedlings. But, already after a few years, the cocoa trees become taller than the other plants and trees, meaning the other species cannot survive in the intensified cocoa farming system. The main agricultural practices applied on farm comprise of activities such as pruning, cutting of mistletoes, spraying of fertilizers, pesticides, and fungicides. Whereas pruning and cutting can be done with machetes or cutlasses, the spraying equipment is capital-intensive and often not available for farmers.

There are two harvesting seasons for cocoa, the main and the intermediate crop season (c.f. Gayi and Tsowou 2016, p. 10). The ripe pods are usually harvested manually with a cutlass or machete and then split. There are about 20 to 50 beans in each cocoa pod and one pound of chocolate requires approximately 400 beans (World Cocoa Foundation 2014, p. 4). After harvesting, the postharvest handling contains two major steps, which receive significant attention from the industry side as they are of crucial importance for the development of the beans'

---

[1] Ruf 1995 developed the pioneering model of „cocoa cycle" which establishes an explanatory relation between a forest rent, migration and economic as well as political environments.

flavor: fermenting and drying (World Cocoa Foundation 2014, p. 4). However, the techniques and duration differ considerably from one producing country to the other and are partly the result of marketing policy incentives to the farmers. Mainly either directly done on the farm or in the farmers' villages, the pulp-covered beans are heaped into piles and are covered with banana leaves. The fermentation process lasts for three to seven days and the drying process depends on the prevailing weather conditions. Farmers pack the dry beans into jute bags which they receive from public or private marketing agents or their cooperatives and transport them to the storing facility of the agent to whom they sell their produce. Planting, growing—including plant protection treatment, harvesting and postharvest handlings are the main activities at the local level, i.e. the farm level.

4.2.2   National Level: Cocoa Marketing

Marketing comprises all activities from the purchasing of the beans from the farmers by any type of buying agent to selling them on the export market. After being sold to international buyers, the beans are transported to processing plants either directly in the producing or in the importing countries (Gayi and Tsowou 2016, p. 11) such as the Netherlands, Germany, or the United States. Which actors are involved in the whole marketing process of the beans, including the sourcing, storing and transporting activities, depends on the marketing strategy of each producer country. However, generally there are two parts of the marketing; the internal and external marketing systems. Internal marketing refers to all pur-chasing, transporting, storage, and documentation activities inside the country as well as quality measurements before export. External marketing generally refers to the selling of the beans to the export market whereby it depends on the con-tract form which party is in charge of shipment and other logistics. Historically, in West Africa, national marketing boards managed both internal and external commercialization of cocoa (Vellema et al. 2016; Kolavalli and Vigneri 2019). However, in the scope of the World Bank's and the IMF's Structural Adjustment Programs all major African cocoa-producing countries introduced liberalization reforms of their cocoa sectors. An UNCTAD (2008) study categorizes the cocoa market liberalization of the four main African producer countries as follows:

- Nigeria: rapid disengagement of the state in 1986/1987;
- Cameroon: progressive disengagement of the state within three stages (1989/ 1990, 1991, and 1994/1995);

- Côte d'Ivoire progressive disengagement of the state in 1994/1995 and in 1999;
- Ghana: disengagement only selectively in 1992/1993 (UNCTAD 2008, p. 3); reforms known as partial liberalization of Ghana's cocoa market.

The difference is that in Ghana only the internal marketing system have been liberalized and Licensed Buying Companies (LBCs) conduct the sourcing of the beans. In the other three countries, regional cocoa merchants—which in many cases are cocoa processors or chocolate manufacturers themselves or their subsidiaries—buy cocoa from middlemen and further sell them to exporters and to a very small extent to domestic processors. Thus, it depends on the sector strategy of each producing country by which stakeholders' local depots are managed or belong to, whether transporting is done by haulers or by the buying companies themselves or whether cooperatives have a stake in the local marketing system. However, even if often neglected in cocoa value chain analyses, because of its relevance for a coherent traceability system, this segment has gained interest from international market actors. The present study will therefore depict the structure of the Ghanaian cocoa marketing system in detail in chapter 5.

4.2.3 Global Level: From Transformation to Retailing

Processing
Traditionally, after the beans have been sold to the export market, they were directly shipped to the importing countries for processing and only a marginal share remained in the origin countries for local processing and consumption. During the 1990s, major chocolate manufacturers abandoned indoor processing and concentrated on product specification. In the same period, a small number of influential TNCs begun to establish processing capacities in the producing countries. The two multinational agro-traders ADM and Cargill, at that point new entrants into cocoa trading and processing, together with Barry Callebaut took over already existing grinding plants or built new ones, for instance in Côte d'Ivoire, Malaysia and Indonesia (Araujo Bonjean and Brun 2016, p. 344; ACET 2014, p. 14). These companies progressively pursue strategies of backwards integration of their business. Furthermore, governments of the producing countries, in West Africa particularly Côte d'Ivoire and Ghana, offer tax incentives to major processors. As a result, mainly transnational cocoa processing companies recently increased their processing capacities in most of the producing countries. Thereby the local processing capacities have rapidly increased, in Ghana by 17% and in

Côte d'Ivoire by 5% per annum, making Côte d'Ivoire the world's largest processing country. In 2013, 86% of the Ivorian grinding capacities were owned by five TNCs (Araujo Bonjean and Brun 2016, p. 344). As Gayi and Tsowou (2016, pp. 7–8) point out, there is a low share of origin countries companies' contribution to these high numbers of processing in producing countries, hence, most of the value is captured by TNCs. Besides, the high-technology plants would generate only limited demand for local employment. Figure 4.3 gives an overview of the main processing countries worldwide. Thus, the rapid improvement of Côte d'Ivoire's position as world's largest cocoa processor does not automatically imply that the country really benefits from this development.

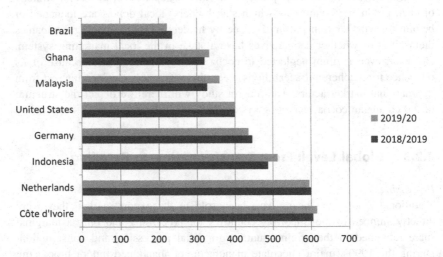

Figure 4.3 Global leading countries of cocoa bean processing in 2018/2019 and 2019/2020 (in 1,000 tons). (Source: Statista 2020c)

There are several intermediary stages of the processing chain from cocoa to the end product, a packed chocolate confectionery available at the store. The main steps are roasting, grinding, pressing, and chocolate making. After the beans are cleaned, they are shelled and roasted—either as a whole or crushed into nibs. Following this, the nibs are then ground into paste also called cocoa liquor. Now there are several options of usage. The liquor can be employed directly as an ingredient for chocolate production or further processed by pressing into cocoa butter and presscake. The cake often will be ground into cocoa powder which, depending from the percentage of fat content, will be used for a wide range of products such as drinking chocolate or bakery powders (Gayi and Tsowou 2016, p. 12; World Cocoa Foundation 2014, p. 5). The cocoa liquor and butter is mixed with other ingredients like sugar or milk powder into a chocolate dough out of which industrial chocolate called couverture is made (UNCTAD 2008, p. 7). Couverture is then sold to chocolate manufacturers which use it as the basis for the production of their wide chocolate ranges. Currently, 70–80% of the world's couverture is produced by only two main processors, Barry Callebaut and Cargill (Fountain and Hütz-Adams 2015, p. 6). Both companies are increasingly vertically integrated and own a number of chocolate brands, too.

Chocolate manufacturing
GCCC analyses do not always differentiate the single production steps in the same way. Particularly the production of couverture sometimes falls under the processing sometimes under chocolate manufacturing segments. For this study, it is more appropriate to classify it as processing as it is done by the world's two largest processing companies. Chocolate manufacturing comprises production steps such as "coating, layering and cutting, and other shaping techniques" (UNCTAD 2008, p. 8). In 2019, the five leading chocolate manufacturers were Mars Inc (USA), Ferrero Group (Luxembourg / Italy), Mondelēz International (USA), Meiji Co Ltd (Japan), and Hershey Co (USA) (International Cocoa Organization 2020c).

Retailing
About 45% of all chocolate confectionery products are bought in supermarkets and discount stores (KPMG 2012, p. 6). The consolidation within the retail sector is known for being high end ever increasing (Fuchs et al. 2009). The commercial marketing and retailing of chocolate products are a core business for retailers. Some retailers like the German grocer Lidl even introduced their own chocolate brands and opened processing plants in some producer countries. Even if retailers generally possess the strongest bargaining positions in global food chains (Fuchs et al. 2009), due to consumers' attachment to brand chocolate products huge chocolate manufacturers hold a considerably strong bargaining position vis-à-vis the retailers (Hütz-Adams 2012, p. 23). These strong positions of retailers and brand manufacturers are reflected by the figures of value distribution (see section "economic pillar" in subchapter 4.4). Considering the small number of companies in the processing segment and their share in the value distribution (see section 4.4.2), their importance also becomes more evident.

4.3   The Hybridization of Governance in the Global Cocoa-Chocolate Chain

The governance of the GCCC is characterized by a multitude of intertwined industrial and global governance developments which shape the functioning of the chain and lead to new forms of control and consent. The above-described most prominent feature of the GCCC, the strongly dispersed local production by millions of small-scale producers and the very concentrated downstream segments dominated by a small number of TNCs, play out a particular dynamic that Fold and Neilson (2016) called the "hybridization of cocoa-chocolate value chain regulation". In this context, the term hybridization refers to the process of an increasing mix of chain-internal and -external governance activities conducted by both, manufacturing and processing lead firms (Fold and Neilson 2016). Accordingly, and as already discussed in subchapter 3.1, the distinction between internal, i.e. industrial governance, from external governance, established by Gereffi (2001) with the separation of the axes "governance" and "institutions", can be considered as misleading since activities in both axes are closely intertwined and overlapping. Fold and Neilson describe the "complex network of rules, standards, norms and interventions instituted by various private sector and civil society actors" (Fold and Neilson 2016, p. 203) which constitute the governance of the GCCC.

When looking at the industrial side of governance, the GCCC has seen a noticeable increase of the processing segment's importance over the past two decades. While the trading segment lost its role in the chain, some of the main traders entered the processing segment and became powerful players. Simultaneously, processes of consolidation of important processing companies occurred (these processes are depicted in more detail in section 4.4.2) and led to a new power arrangement in the chain that Fold (2002) called a bi-polar buyer-driven chain—in contrast to the former common interpretation of the GCCC as a buyer-driven chain. The term "bi-polar" refers to the division of power between the processing and manufacturing segments. The entry of large agro-traders which purchase and transform many different commodities resulted in a rapid increase in logistical efficiency in the chain, particularly regarding transport and storage. This is because the large agro-food TNCs transfer their technical, organizational, and managerial competencies from other business lines to the GCCC (Fold and Neilson 2016, p. 196). Yet, due to this distinct power division between the two transformation segments, there is almost no complete vertical integration in the chain. Nevertheless, recently, there is a strong trend towards backwards integration by the dominant processing companies into the local marketing and exporting activities. The way control and backwards integration is achieved is different at each producing-country level and depends on the respective sector policies. In many cases TNCs from the processing segment operate through their subsidiary firms or have arrangements with local traders. Fold and Neilson describe the motivation of activities to increase control at the production and local levels as a result of the industry's increasing concerns over sufficient quality and quantity of beans. This is in line with the above described apparent double pressure which led to the rapid spread of CSR in the decades after the turn of the millennium. For instance, as will be further argued and shown below, CSR—mainly in the form of third-party sustainability certification—functions as an important tool to support that strategy of backwards integration and improved control over local production and marketing activities in the GCCC.

In the course of their CSR strategies, mainly the large processors established direct relations with suppliers in producing countries and implement sustainability standards schemes at the local level. Running a sustainability project implies traceability whereby the information about the beans' origin and improved knowledge on qualities is held by the implementers. Linked to these programs, lead firms also participate in cocoa sector policy platforms in producing countries where a lot of local information is circulated and sometimes act in cooperation with the public sector. These new activities linked to sustainability interventions improve the competitiveness of a company.

Chain-external governance activities at the global level of the GCCC mainly comprise efforts to participate in institution-building processes or in the development of sector-relevant standards and norms. Since the early 2000s, the relevant institutions in the GCCC have had to respond to consumers' increasing demand for fair and child labor free chocolate products, and more recently also for more ecological production practices, as a cocoa production free of deforestation. While CSR became a major tool for TNCs from the GCCC it also became an integral concept in most of the cocoa agendas and served as a legitimization for the increased participation of TNCs at this level. Here, it is not only brand manufacturers which were initially mostly targeted by consumers' critiques that are involved, but also cocoa processors take increasingly part in sector deliberations, reflecting the general trend of opening up governance rooms to the private sector. As discussed in the theoretical chapter before, the active participation in public-private or private-private international and transnational institutions is an opportunity to shape sector dynamics and policy processes proactively. Cocoa sector projects, product and process standards for the cocoa-chocolate chain or other policy tools and guidelines are developed in a number of institutions of different composition and scope. The most important ones are presented here and a particular emphasis is given to the new role the private sector and CSR play in them.

The International Cocoa Organization (ICCO)
The most important institution for international cocoa-chocolate regulation, gathering and dissemination of information and research findings on cocoa is the International Cocoa Organization (ICCO). It is an intergovernmental body where almost all producing and consuming countries hold a membership. Under the auspices of the United Nations Conference on Trade and Development (UNCTAD), ICCO was established in 1973 with the aim to implement the first International Cocoa Agreement (ICA). The main objective was the prevention of excessive price fluctuations. Two main tools to achieve this objective were the introduction of export quotas and, being located in London, holding cocoa buffer-stocks to regulate the market (cf. Fold and Neilson 2016). However, being shaped by the general liberalization trends from the 1980s onwards, this public stabilization system was abandoned stepwise. With the fifths ICA, member countries agreed on reducing ICCO's tasks. They decided to implement a reporting scheme on their cocoa stocks as well as their exports and imports. Thereby, the responsibilities of the ICCO were reduced to the monitoring and publication of market developments such as the daily cocoa price.

With the liberalization of the cocoa sector, ICCO lost its regulatory importance but remained an important platform for member country dialogue on pressing industry challenges (United Nations 1993). ICCO also runs a couple of projects in the fields of quality improvement, fine flavor cocoa, cocoa marketing and trade or price risk management. ICCO's highest decision-making body, the International Cocoa Council, has equal voting shares for producing and consuming member country representatives. It meets twice a year and is composed of member countries' representatives. The council defines strategic measures, such as financial and programmatic plans, and oversees their implementation.

Importantly, ICOO's decisions are based on advice from its four subsidiaries, the Economics Committee, the Administration and Finance Committee, the Expert Working Group on Stocks, and the Consultative Board on the World Cocoa Economy. The latter plays an important role for the participation of the private sector. It already exists since 2003 but an extended version was affirmed in 2010 in the frame of the seventh ICA, negotiated in 2010 in Geneva and put into force 2012. Next to its mandate to strive for a Sustainable World Cocoa Economy, the Consultative Board is seen as an "important breakthrough" (International Cocoa Organization 2020a) of the ICCO which was established for the "active participation of experts from the private sector in the work of the Organization and to promote a continuous dialogue among experts from the public and private sectors" (International Cocoa Organization 2020b). Its members are representatives from brand manufacturers and processors, industry associations, public-private partnership organizations such as the German Initiative on Sustainable Cocoa (GISCO) or the Dutch third-party certification body UTZ, and also purely public institutions such as the Ghana Cocoa Marketing Board (COCOBOD) and the Ivorian Conseil Café Cacao (CCC). Through the active participation in this central advisory board at the international level, the private sector has a close link to the decision making organ of the ICCO. The Consultative Board is therefore clearly an important tool for TNCs to influence institutional constellations surrounding the GCCC.

The World Cocoa Foundation
Another important institution in the GCCC is the purely private sector membership organization World Cocoa Foundation (WCF). It represents more than 80% of the GCCC industry with over 100 companies from all branches along the GCCC, namely "farm-level input providers, financial institutions, cocoa processors, chocolate makers and manufacturers, farmer cooperatives, cocoa trading companies, ports, warehousing companies, and retailers" (World Cocoa Foundation n.d.c). Founded in 2000 with an initial focus on the development of improved plant material, the organization became the main "meta-governance body" (Bitzer et al. 2012, p. 369) which provides a steering forum for industry coordination and bundles technical, financial, and knowledge resources from industry members. Members cooperate in their pre-competitive fields, most notably the enhancement of farmers' capacities or the general improvement of the industry's legitimacy (cf. Bitzer et al. 2012). In this light, the WCF declares that it's "...vision is a sustainable and thriving cocoa sector—where farmers prosper, cocoa-growing communities are empowered, human rights are respected, and the environment is conserved" (World Cocoa Foundation n.d.d).

Based on this, it states its mission as follows:

"Our mission is to catalyze public-private action to accelerate cocoa sustainability. We champion multi-stakeholder partnerships, aligned public and private investment, policy dialogue, and joint learning and knowledge sharing to achieve transformative change in the cocoa supply chain." (World Cocoa Foundation n.d.d)

One can see how the discourse of industry's responsibility for improved sustainability in the GCCC is applied here. WCF proactively uses the concept of TNCs' responsibility for improved sustainability in the chain to justify the expansion of its members to several fields of external governance, including policy deliberation and local development activities.

Apart from shaping the discourse, the WCF itself acts as a main industry representative in policy deliberations at both national and international levels and often serves as a host of corresponding activities. In addition, next to policy dialogue and the development of an aligned industry strategy for the cocoa sector (CocoaAction), WCF gives funds to its members on a competitive basis which beneficiaries can use to implement projects on the ground. The WCF also implements sustainability programs itself. Since 2010, three main programs have been run: The Cocoa Livelihoods Program, the African Cocoa Initiative, and, since 2016, the Climate Smart Cocoa Program. The programs seek to advance industry initiatives at the local level of cocoa production. Important funders are the Bill and Melinda Gates Foundation, the Walmart Foundation, and USAID. Implementing member companies often cooperate with development agencies such as the German agency for development cooperation (Deutsche Gesellschaft für Internationale Zusammenarbeit, GIZ) or international NGOs such as CARE International. The WCF has its headquarters in Washington but also runs national offices in Accra and Abidjan. Working through local staff and being closer to the target areas allows a better coordination and of programs' implementation. (Interview with a WCF representative in Accra, 2015)

International Cocoa Initiative

The industry-led International Cocoa Initiative (ICI) was established in 2001 and was an immediate result of the Harkin-Engels protocol against child labor in the GCCC which will be described below in section 4.4.3 on the social sustainability pillar. Its main mandate is seen as the promotion of "child-protection in cocoa-growing communities and to ensure a better future for children and their families" (International Cocoa Initiative n.d.a).

Until today, most of the lead firms in the GCCC are members of the ICI. The board consists of the three most important processors, Barry Callebaut, Cargill, and Ecom, as well as the leading chocolate manufacturers Mondelez International, Hershey, Mars Incorporated, and Ferrero (International Cocoa Initiative n.d.b). During the 2000s, the initiative was an important catalyzer for the rapid spread of anti-child labor projects in Côte d'Ivoire and Ghana which TNCs had started to implement in the frame of their CSR projects. While the initial focus was supposed to be on livelihood improvement and anti-child labor measures, these projects, often implemented in partnerships as for instance with the Sustainable Trees Crop Program (STCP) conducted by the International Institute of Tropical Agriculture (IITA) (cf. Fold and Neilson 2016), quickly turned to emphasize technical sustainability approaches which were more concerned with production sustainability than with a holistic improvement of cocoa producers' living conditions. Today the ICI's role in the GCCC is less important and as of 2020, ICI only runs three community projects. The organization presents its key activities in sustainability governance as follows:

"We participate in different international meetings with civil society, the cocoa indus-try, cocoa-producing, and cocoa consuming governments. Serving as a voice for chil-dren in the global agenda, we influence international debates, policies and strategies for tackling child labour in cocoa and promoting child rights." (International Cocoa Initiative n.d.b)

As can be seen, being a purely industry body, the initiative presents itself as an independent child rights advocator which strives for sustainability in the GCCC. The systemic reasons for child labor, however, such as farmers' poverty and its sources, are completely out of focus in such an industry body.

Partnerships and platforms
Since the 2000s, many different forms of partnerships have been implemented along the search to improve sustainability in the chain. Partnerships exist at all administrative levels and between a large number of diverse stakeholders. There are purely industry partnerships, but public-private partnerships dominate. Next to TNCs, international financial institutions, private donors such as the Bill and Melinda Gates Foundation, national development agencies, UN bodies (UNDP, FAO), international and local NGOs, Think Tanks, universities and other research institutions, public authorities, and farmers' cooperatives come into different part-nership arrangements in order to tackle the apparent joint objective of cocoa sustainability. Important partnerships which have been implemented in the pro-ducing countries are between the STCP and the Conseil Café Cacao Côte d'Ivoire or in Ghana between the WCF, GIZ and STCP (Muilerman and Vellema 2017, p. 88). In Ghana, for instance, the UNDP initiated a National Cocoa Platforms which was then handed over to COCOBOD and served as a platform for sec-tor policy dialogue between COCOBOD and a number of TNCs, certification bodies and development agencies. In the importing countries, there are many ini-tiatives to support cocoa sustainability. Exemplarily, the multi-stakeholder ISCOs (Initiatives on Sustainable Cocoa) in a number of European countries such as the Dutch Initiative on Sustainable Cocoa, the French Initiative on Sustainable Cocoa (FRISCO) or the German Initiative on Sustainable Cocoa (GISCO). can be mentioned.

The sustainability certification programs of Rainforest Alliance/UTZ and Fairtrade Labelling Organization International

As outlined above, over the past decade, sustainability or fair trade certification schemes advanced to important tools for risk management and private sustainability governance in the GCCC. Before the merger of the Dutch UTZ and the US Rainforest Alliance in January 2018, these two organizations together with Fairtrade Labelling Organization International (FLO) ran the most important sustainability certification programs in the GCCC. The consolidation of the two most relevant certification programs for cocoa is important not only from a governmental perspective but also for the development of the present study since the empirical part analyses the implementation of one UTZ certification project in Ghana over the period 2015 to 2017, prior to the merger. Therefore, the fusion will be described here but after this, the two organizations are still described as separately.

The major aim of the merger between the two biggest certification programs was to become more efficient. The new organization uses the name and seal of Rainforest Alliance but the merger is communicated as a fusion of two equally strong partners which seek to pool their knowledge and strengths to respond efficiently to increasing challenges of sustainability. From the new Rainforest Alliance self-description, strengths of the old Rainforest Alliance lay in their strong market position and high acceptance by consumers as well as the expertise regarding conservation of landscapes. The core competencies of UTZ are described as being located in the fields of business, sector engagement, and the development of value chains. The new organization states that keeping the best criteria from both former standards and transferring them to a joint program within a consultative process led to the development of a new certification system. This implies new standards for producers and a new chain-of-custody system. The organization sees its role in bringing together all kind of stakeholders from the respective sectors it is working in to foster dialogue on sustainability (Rainforest Alliance 2020a).

Generally, certification programs work based on producers' application and compliance of the standard requirements set by the respective certification organization. While there are different systems of delivering the individual norm catalogue to producers, independent certification bodies cooperate with the standard-setting organizations and conduct on-farm inspections to monitor and certify compliance. Often subsumed either as sustainability or fair trade labels, the UTZ, Rainforest Alliance, and FLO have quite distinct focuses. FLO is mainly concerned with minimum prices, living wages, pre-financing of harvest periods, and a fair trade system. Rainforest Alliance, before the merger, was mostly concerned with ecological production, the protection of biodiversity, and education on environment and ecological farming practices. For UTZ, the focus was mainly on good agricultural practices, social standards, and traceability of produces.[2] The German consumer safety group "Stiftung Warentest" in 2016 published a comparison of many sustainability seals. Comparing the three mentioned certification programs, it evaluated their performance concerning general requirement levels, implementation into practice, management of the organization, and other features such as the payment of minimum prices or premiums. Table 4.1 summarizes these findings.

Table 4.1 Comparison of the three sustainability seals by "Stiftung Warentest"

Topic	FLO	Rainforest Alliance	UTZ
Requirement level	Good	Moderate	Moderate
Implementation into praxis	Very good	Good	Good
Management of the organization	Very good	Good	Very good
Existence of minimum price	Yes	No	No
Payment of premiums	Yes	No	Yes

Source: Stiftung Warentest 2016.

The empirical part of this study focusing on the implementation process of one large cocoa certification program as the core CSR strategy of one of the main cocoa processing companies, it is important to bring attention to lead firms' participation in the development of some of the sustainability standards which are used as normative references during the implementation of sustainability programs. The former certification program UTZ, which was "the largest program in the world for sustainable cocoa" (UTZ—Rainforest Alliance 2020b) developed the standards which the farmers are supposed to fulfill on a multi-stakeholder basis where lead firms were participating.[3] Its twelve members-counting Standards Committee consisted of six industry representatives (among others from Cargill, Ferrero, Dole and Rewe), three civil society representatives, one government agency representative, one certification bodies' representative as well as

---

[2] For a good overview see Coffee Circle 2020.

[3] In an interview with a Dutch representative of a major transnational processing company, the interviewee stated that his company had a driving seat in the development of the UTZ standards. However, this statement could not be verified by the author, but it might give an indication on internal dynamics of private sustainability standard developments.

one member of Rainforest Alliance. Similarly, the boards of directors of the three organizations have a high representation of members with long histories of key positions in TNCs mainly from food and retail but also other sectors.

4.4   Key Challenges to Achieve Sustainability in the Global Cocoa-Chocolate Chain

While many governmental constellations at different levels of the GCCC are in place seeking to tackle most pressing sustainability challenges, sources are deep-rooted, mostly structurally caused, and often neglected at the presented fora and institutions. Challenges concern all three pillars of sustainability, namely environmental, economic, and social aspects and are strongly entangled. Hence, the picture of the whole sector sustainability is highly complex and each pillar itself and the linkages between them could easily be the target of a separate study. Accordingly, the following subchapter can only introduce the some of the key challenges. For each sustainability pillar, one or two of the core deficiencies are presented, particularly those which are most often discursively linked to the introduction of CSR in the GCCC.

4.4.1   Environmental Pillar: Focus on Deforestation and Biodiversity Degradation

As it counts for all major agricultural commodities, cocoa production has a large ecological foot print. It has contributed to massive landscape changes in the tropical forest belt. The expansion of cocoa plantations into virgin forest lands has been identified as one of the major drivers of the immense deforestation over the past 20 years (Asare et al. 2014; Ameyaw et al. 2018; Schroth et al. 2016; Noble 2017). According to Gockowski and Sonwa (2011), the Guinean rain forest of West Africa has reduced to $113{,}000km^2$, what is only 18% of its original size. This is not alone due to cocoa plantations but also other agricultural cultivation such as the plantation of cassava, oil palm, and plantain (Gockowski and Sonwa 2011) or logging. However, a report from the Tropical Commodity Coalition (2008, p. 8) indicates that cocoa is responsible for approximately $80{,}000km^2$ loss of the world's tropical forests. Forest degradation at such levels is accompanied with major ecological problems such as high greenhouse gas emissions, the loss of biodiversity, and soil erosion. As Gockowski and Sonwa (2011) point out, the tropical forest conversion into agricultural lands releases the biggest share of CO_2

emissions due to land changes and is responsible for an estimated loss of 27,000 species each year (Wilson 2002). Moreover, soil erosion in cocoa plantations is regarded to be high because of cocoa's thick leaves which are slow to decompose thereby suppressing other vegetation (TCC 2008, p. 8).

In West Africa, the contribution of cocoa to deforestation is mainly attributed to large migration movements into virgin forest zones. Cocoa farmers from other regions or people from all over the country or even from neighbouring countries looking for a new livelihood opportunity settled into wet evergreen forest zones (Asare et al. 2014) in order to benefit from the high fertility of forest soils, what Ruf (1995) has called "forest rent" (in more detail in chapter 5). Settling in the forest areas, farmers used traditional cocoa production practices. The forest canopy was thinned but not cut off completely and cocoa seedlings were planted under the remaining canopy as a second stratum. During the early period of cocoa expansion, a multi-strata cocoa agroforest system was predominant. Ruf (2011, p. 375) categorizes cocoa plantation schemes as follows: complex agroforests with high levels of biodiversity what has been the old type of cocoa cultivation, cocoa agroforests, simple agroforests, and full-sun systems which establish a monoculture of cocoa trees. Until the 1990s, most cocoa cultivation was based on extensive farming in a cocoa agroforest setting, mainly without considerable application of fertilizers and other agrochemical inputs. Next to its immediate role for deforestation, extensive cultivation has been highlighted as the main reason for farmers' low productivity. Asare et al. (2014) identify a large gap between the actual yields in the 1990s and the possible output under intensification scenarios. With traditional extensive practices, farmers may achieve approximately 400kg/ha whereas their potential yields applying new technologies and chemical inputs would rise to more than 1000kg/ha (Asare et al. 2014). According to production scenario modelling by Gockowski and Sonwa (2011), an intensified full-sun cocoa plantation would yield a 148% to 161% higher return than extensive shade systems. From the 2000s onwards, various extension agencies begun to engage in the provision of technical advice to cocoa farmers for the application of intensified cocoa technologies, mainly understood as Good Agricultural Practices (GAP). These techniques comprise of improved planting material namely hybrid cocoa, the application of pesticides, fertilizer and herbicides, and weed control and pruning. There is a direct link between the planting of hybrids which are more sun-resistant and the establishment of shade-free plantations which are monoculture systems (Ruf 2011; 2007). In such a full-sun system established on former rainforest land, only one level of canopy storage remains, that is the cocoa trees, and all large trees are removed. Up until the mid-2010s, the mainstream assumption was that, economically speaking, there is a negative correlation between cocoa productivity and species richness due to the competition for water, sunlight, and other nutrients (Gockowski and Sonwa 2011). On such intensified farms and with the proper use of agrochemical inputs and agricultural practices, farmers can expect much higher yields for a period of 25 to 30 years. Ruf (2007) describes the "new Ghana cocoa boom" which occurred in the country during the 2000s. Contrary to the former production increases mainly based on forest clearing, a huge production growth occurred which this time was mainly stemming from the introduction of intensification technology; what he calls a *green revolution in the Ghanaian cocoa sector* (Ruf 2007). Mainly distributed by governmental or international development agencies as well as by the private sector in the course of the by then introduced diverse CSR activities, the new kind of hybrid seedlings quickly spread among farmers. Farmers tended to be eager to adapt the new plant material which is attributed to higher and quicker returns and transferred the notion of being a modern farmer (Ruf 2011). Particularly on the latest cocoa pioneer fronts in West Africa's high forest zones, highly productive cocoa farms have been established which are full-sun systems without a relevant number of shade trees remaining.

However, over the past few years, a U-turn in agricultural advice occurred and the importance of keeping farm diversity for both farmers' income and biodiversity restauration is now largely acknowledged. Already prior to this recent change, there has been an ongoing literature debate on the desirability of intensified cocoa production. The trade-off between forest land protection through agricultural intensification and the much higher biodiversity loss on monoculture small-holder farms in the rainforest zones has been increasingly recognized. While intensification and the shift to full-sun schemes was considered to have a strong positive impact on yields and reduces forest conversions on an immediate basis, concerns are raised regarding the long-term sustainability of these systems. The loss of biodiversity on full-sun farms is accompanied with a higher vulnerability to cocoa pests and diseases with a steady dependency from chemical inputs. Franzen et al. (2007) make the point that the picture of high economic and environmental profitability of full-sun farms might be also unclear because the real high productivity of these hybrids goes drastically down after a shorter term. In addition, since hybrids are strictly dependent from chemical inputs, new environmental challenges might easily arise in areas which are completely altered into shade-free zones.

Due to increased awareness of the importance of environmental sustainability, ecological concerns are much more present in cocoa policy debates than before and have become a business strategy, too. There is a new consensus among public and private cocoa actors that shade trees have to be reintroduced into the cocoa farming systems in order to protect the soil and gain back biodiversity (Yamoah et al. 2020). The ecological services of agroforestry systems are increasingly recognized and concepts such as *agroforestry intensification* are promoted (Gockowski and Sonwa 2011). How feasible and ecologically sustainable such approaches are remains to be observed in future, particularly under changing climatic conditions in the region. Rainfall patterns are drastically changing in the West African high forest zones. The cocoa-growing suitable humid climate is disappearing due to the decrease of rainfalls in the area Codjoe et al. (2013).

Climate-smart cocoa strategies are increasingly integrated into policy deliberations and sector strategies coordinated, for instance with the World Cocoa Foundation's Climate Smart Cocoa Program. Rainforest Alliance in cooperation with WCF and the International Center for Tropical Agriculture developed a training manual on climate-smart agriculture in cocoa (Rainforest Alliance 2018). Indeed, climate-smart cocoa requires improved coordination and integration of different stakeholders' efforts. For example, improved landscape planning and data management to be combined with existing strategies of production increase (Asare et al. 2014). The industry is getting ready to face environmental challenges since these are one pillar of the three main sustainability threats of cocoa production. While the forest rent is getting to its end, intensification and full-sun schemes tend to be more vulnerable to climate change impacts such as reduced rainfalls and even aggravate them. It will be important to follow these dynamics and how TNCs respond to them. Under the current conditions, it is highly likely that climate-smart agriculture is combined with the introduction of new technologies at the farming level. In this course of the cocoa farming digitization process, some lead firms will likely become the owners of most crucial farming data and technologies. This process might tend to foster the already existing patterns of monopolization and bargaining inequalities in the industry, a trend which will be described in the next section.

4.4.2 Economic Pillar: Focus on Value Distribution and Power Concentration in the Global Cocoa-Chocolate Chain

The GCCC is characterized by a strong asymmetry in value distribution. The largest share (44.2%) of value added of a bar of chocolate goes to the retail sector and also a high share to the manufacturing segment (35.2%). The processing companies make 7.6% whereas 4.2% of the whole value go to public institutions in the producing countries. 2.1% are generated by the local purchasing and transport segment and only 6.6% go to about six million small-scale farmers in all producing countries (Fountain and Hütz-Adams 2015). Figure 4.4 shows the shares of value distribution between the different stakeholders in the chain in the year 2014.

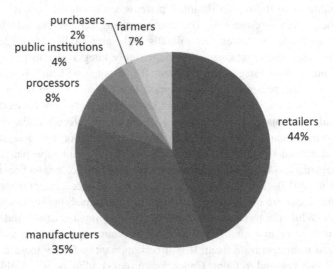

Figure 4.4 Value Distribution in the Global-Cocoa Chocolate Value Chain. (Source: own elaboration based on INKOTA 2016)

During the post-liberalization period from 1999 to 2010, the farm-gate price ratio fell about 10% compared to the 1985–1998 period (Araujo Bonjean and Brun 2016, p. 346). In the world's two main producing countries, the small-holder farmers only have a mean daily income of 0.5 USD (Côte d'Ivoire) and of 0.84 USD (Ghana) (Fountain and Hütz-Adams 2015, p. 0), what is far below the World Bank's International Poverty Line of $1.90. Living under the poverty line, for many cocoa farmers implies a constant deprivation from their economic/ financial capital negatively impacting their human, social, and natural capitals, too. Families living in such unsustainable rural livelihoods are often faced with food insecurity, lack of access to health services, children unable to go to school and their active participation in social activities is strongly limited.

Given that the major retailers, brand manufacturers, and processors are based in the USA or Europe, the main value added in the chain is captured outside of the producing countries. Accordingly, the paramount shares of profit of the chocolate industry with a total value of 103.28 billion USD in 2017 (Globe-Newswire n.d.) is made in industrialized and emerging economies. In 2008/2009, 11 billion USD of the by then 87 billion USD industry remained in Africa with $6 billion counting for cocoa beans sales and $5 billion for semi-processed products (ACET 2014, p. 11). The World Bank (2008) states a decline of producing countries' share of retained value from abound 60% in 1970–72 to about 28% in 1998–2000 (cited after Abdulsamad et al. 2015, p. 2). The joint market share of the top five chocolate companies accounts for more than half (55%) of the global chocolate market:(the US Mars Incorporated makes 14.4% and the US Mondelez International—which is a merger of the former food and confectionery giants Kraft and Cadbury—13.7%, the Swiss Nestlé 10.2%, the Italian Ferrero 9.5% and the US Hershey 7.2%, (Statista 2017, p. 20). Figure 4.5 shows the shares of chocolate manufacturing companies in the global chocolate market.

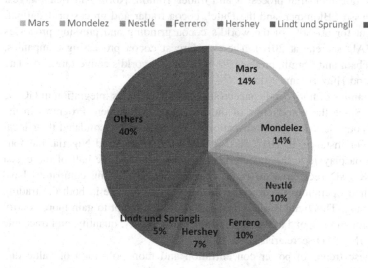

Figure 4.5 World's leading chocolate manufacturing companies. (Source: own elaboration based on Statista 2017)

An even higher degree of consolidation was witnessed in the cocoa processing segment and affects the economic sustainability of the GCCC. The newly-entered US food processing and commodities trading giants Archer Daniels Midland (ADM) and Cargill acquired many European processors and rapidly gained large market shares during the 1990s (Araujo Bonjean and Brun 2016, 347 ff.). Subsequently, the processing segment has seen many takeovers during the last decades. The Swiss company ECOM Agroindustrial is a global commodity trading and processing company and acquired the cocoa business arm of the London-based commodity investment firm Armajaro which became known for its aggressive buying-up strategy—in 2010 it was responsible for 7% of world's cocoa purchases (Araujo Bonjean and Brun 2016, p. 348). The Singapore-based agribusiness company Olam International acquired the local cocoa purchasing arm of ADM, whereas Cargill, one of the world's largest agricultural commodity traders, bought ADM's processing capacities in Europe. The Swiss cocoa trader and processor as well as chocolate manufacturer, Barry Callebaut, acquired the cocoa ingredients division of Petra Foods Limited and is now the largest chocolate and cocoa product supplier in the world (Barry Callebaut 2020). Except for Barry Callebaut, which only works with cocoa, these companies are transnational agro-food companies which invest in many crops around the world. Together with the French agro-industrial processor and trader Touton, North America's largest cocoa processor, Blommer, and the Dutch cocoa trader and processor, Continaf, they account for 60–80% of the world's cocoa grinding and pressing processes (cf. UNCTAD secretariat 2016). The two largest cocoa processing companies, Barry Callebaut and Cargill, produce 70–80% of the world's couverture chocolate (Fountain and Hütz-Adams 2015, p. 6).

The dynamics of horizontal concentration and backwards integration in GCCC are strong. Since the introduction of the Structural Adjustment Programs in the producing countries, powerful transnational corporations consolidated their local positions. For instance, in 2012, in Côte d'Ivoire, Ghana and Nigeria, the four leading cocoa players in these countries bought more than the half of the cocoa beans (UNCTAD secretariat 2016, p. 2). All main processing companies have extended their operations down to the farm level and engage in both the trading and processing. This trend is often interpreted as an attempt to gain more control in their chain and respond to the demand for more quality, quantity, and traceable products (UNCTAD secretariat 2016, p. 8).

How these trends of power concentration and monopolization of value distribution affect the functioning of the chain and particularly the economic sustainability of cocoa farmers and producing countries is debated controversially in the literature. Recent studies investigate the effects of this oligopolistic structure of the GCCC, where a large number of (largely unorganized) sellers (cocoa farmers) are positioned vis-à-vis a significantly smaller number of buyers. Research particularly focuses on price-setting mechanisms, competitiveness among the players, efficiency gains, consumer benefits, as well as the effects in the producing countries and on farmers (cf. Araujo Bonjean and Brun 2016; Cappelle 2008; Oomes et al. 2016; UNCTAD secretariat 2016; UNCTAD 2008).

However, findings of these studies are inconclusive, most notably on the effects in the producing countries. There seems to be a general consent that benefits of cost savings occur due to efficiency gains due to economies of scale and of scope (Oomes et al. 2016, p. 9). Furthermore, some authors agree that there is no hard evidence of price agreements amongst the lead firms which would affect the level of the farm-gate prices and thereby directly contribute to farmers' poverty. Rather, the possibility that some of the cost savings are passed onto consumers is mentioned but the likelihood that cost savings are also passed on to farmers is seen to be of greater doubt (Traoré 2009). Cappelle (2008) expresses optimism that a higher degree of consolidation could help to improve sustainability in the chain because of lead firms' central position and the simplification of multi-stakeholder activities due to the reduced number of players. Similarly, Oomes et al. (2016) do not see the trend of power concentration as causing lower farm gate prices but argue that the poverty of cocoa farmers stems from their low productivity and the lack of income diversification opportunities. In their view, sustainability of the chain could be achieved through a double strategy which assists public agencies during the creation of income alternatives and also supports the remaining farmers to increase their productivity. In contrast, Araujo Bonjean and Brun (2016) suggest that the high degree of consolidation might lead to stronger collusive behavior. An UNCTAD study by Gayi and Tsowou (2016) also argues that the concentration of a very small number of powerful buyers worsens the bargaining position of the dispersed small-scale farmers, particularly in completely liberalized cocoa sectors like in Cameroon, Nigeria or Indonesia. Oomes et al. (2016) also recognize the possibility that traders might in some cases abuse their position, especially in remote areas. They highlight that other mechanisms of power exertion than price agreements by strong market players vis-à-vis farmers or cooperatives should also be taken into account. For instance, in Côte d'Ivoire, many cooperatives become dependent on the increasingly smaller number of processors which locally act as buying agents. In that context, processors run pre-finance systems which cooperatives often rely on due to their small margins and lack of access to other credit facilities.

4.4.3   Social Pillar: Focus on Child Labor

At the latest with the British documentary "Slavery: A Global Investigation" by Brain Woods and Kate Blewett of True Vision Production[4], awareness of the problem of child labor in the West African cocoa industry spread among Western consumers. The documentary showed children being trafficked from Mali and Burkina Faso to work on Ivorian and other West African cocoa farms. The problem has been already investigated and documented earlier, though. International NGOs such as Anti-Slavery International and International Research Think Tanks like International Institute of Tropical Agriculture (IITA) have been seeking to raise awareness and resolve these issues.

Child labor can be regarded as one of the worst results of the high-poverty rate among cocoa farmers. When farm-gate prices are equal to or less than production costs, in the absence of alternative income opportunities, the response of producers might not be stopping the production but the exploitation of labor and the engagement in environmental damaging practices (cf. Blowfield 2003, p. 17). Therefore, a closer look at the sources and extent of the problem is required. There has to be a clear understanding of what is talked about when discussing child labor in cocoa farming. Not all work can be considered as harmful or exploitative to children. There are two types of child labor in cocoa. One is children working after being trafficked. Children are trafficked between West African countries and sold into different types of economic activities such as domestic servants, agricultural workers, child beggars, or prostitutes (Anti-Slavery International 2004, p. 52). The second aspect of child labor is children helping on their parents' farms; a practice partly linked to tradition and partly to rural household poverty. This practice does not necessarily violate international conventions on child labor. It is important to have the clear distinction of these two forms of child labor in mind for the discussion of child labor in cocoa.

Boas and Huser (2006, p. 10 f.) divide child farm work as follows:

- Family labor: children of the farmer or children of close relatives who live on the farm,
- Foster labor: children with well-established kinship or communal tie to the household,
- Salary labor: children who work for a salary without any kind of family, kinship or communal tie to the farm household in which they work.

---

[4] Online accessible under https://www.youtube.com/watch?v=DEN3o92Zkxk (last accessed 08.07.2018).

"Slavery" is missing and should be added to the listing. Even if international conventions clearly spell out what is to be considered as child labor and what not, there is still room for interpretation that fuels the discussion because different frames of reference are applied. For instance, the UN Convention on the Rights of the Child recognizes in Article 32 the right of the child to be protected from economic exploitation and from performing work that is "likely to be hazardous or to interfere with the child's education, or to be harmful to the child's health or physical, mental, spiritual, moral or social development." (United Nations Human Rights Office of the High Commissioner 1989)

Also the ILO Conventions '182 on the worst forms of child labour', '184 on safety and health in agriculture' and '138 on the minimum age for labour' determine what should be understood as abusive child labor in the context of cocoa farming and what not. Convention 138 states that any economic activity performed under the age of 15 would be child labor. In contrast, the African Charter on the Rights and Welfare of the Child follows the UN Convention and also defines child labor as "any work that is likely to be hazardous or to interfere with the child's physical, mental, spiritual, moral, or social development." (International Law and Policy Institute 2015, p. 12 f.)

Anyhow, the general consent seems to be that carrying out work activities which might be harmful to children's health is not allowed under the age of 18. In cocoa farming, typical work activities where children in a family or labor arrangement are involved are lighter work activities such as clearing the ground, weeding, or gathering cocoa pods and fermenting and drying cocoa beans. Nevertheless, there are also cases reported where children were applying pesticides and spreading fertilizer, pruning, and harvesting, which are more dangerous farming activities (Boas and Huser 2006, p. 10). Finally, it is difficult to assess which activities are actually done by children at the farm level and there is a lack of reliable data. It becomes clear that there are different degrees of child work which have to be considered when assessing the problem. Where some authors argue that work under the age of 18 might qualify as prohibited child labor, even if carried out voluntary (Save the Children Canada 2003), others contend that child labor occurs when children are too young (under the age of 15) or when children under 18 are endangered due to the engagement in hazardous work (Boas and Huser 2006). Hence, the interpretation of what child labor actually is, differs across cultural, legal, and sector-specific factors (International Law and Policy Institute 2015, p. 9).

Anyhow, and importantly, the pictures of the documentary provoked a public outcry and Western governments alike felt urged to rapidly react on the problem. Being blamed to be non-reactive on the problem in their supply chains, major

European and American chocolate manufacturers initially denied their awareness on the problem and their responsibility for conditions in the cocoa farms (Schrage and Ewing 2005, p. 104). One day after the broadcast, the UK Biscuit, Cake and Confectionery Alliance (BCCCA) stated publicly: "we do not believe that the farms visited by the programme are in the least representative of cocoa farming in Côte d'Ivoire, although the claims cannot be ignored." (cited after Anti-Slavery International 2004, p. 54)

Shortly after, an US initiative headed by U.S. Representative Engel to introduce a legislative amendment to the agricultural bill which would require the industry to develop a certification scheme for child labor free chocolate products was approved by the House of Representatives. This has been opposed by the industry and lobby activities for voluntary measures were initiated. As a result, negotiations between the various stakeholder groups begun. Headed by the U.S. Senator Harkin and Engel, in 2001, these negotiations led to the signing of the Harkin–Engel Protocol, also referred to as Cocoa Protocol. It was aiming at ending the worst forms of child labor and forced labor in the cocoa industry (Hütz-Adams 2009, p. 20; Anti-Slavery International 2004). However, the protocol is a voluntary public-private measure and contains only guidelines for producing cocoa in line with the ILO Convention 182. Its importance might be rather located in the structural effects of its six-step action plan:

1. Public statement of need for and terms of an action plan
2. Formation of multi-sectoral advisory groups
3. Signed joint statement on child labour to be witnessed at the ILO
4. Memorandum of cooperation
5. Establishment of joint foundation
6. Building toward credible standards. (Chocolate Manufacturer Association 2001)

Signed by the presidents of the Chocolate Manufacturers Association and the World Cocoa Foundation and witnessed by the US senators Harkin and Kohl, by the Congressman Engel and by the Ivorian Ambassador Bamba, as well as by several NGO representatives, the Harkin-Engel protocol paved the way for the foundation of the private sector-led International Cocoa Initiative (ICI), and many further private and public-private programs against child labor in the sector followed. The protocol can be understood as the starting point of CSR in the GCCC. But is has to be questioned to which degree underlying structural causes of child labor are tackled by the industry-led initiative. In Europe, the campaign

„Make Chocolate Fair!" is supported by 117 member organizations from 16 European countries. The activist movement states that the main reason of child labor is continuing poverty among cocoa farmers and their families due to too small cocoa prices in the world market. The incomes of farmers would have to triple in order to provide a decent living (INKOTA-netzwerk 2020b). The movement urges the industry to assume their responsibilities and to end child labor in their supply chains. However, whether the current approach of CSR in the GCCC is supported and expected to restructure the chain in such a way that cocoa farmers will finally be able to make a decent living is not clearly stated. The following subchapter provides an overview of the most important CSR activities in the GCCC.

4.5   CSR as a Response to Sustainability Deficiencies in the Global Cocoa-Chocolate Chain

> "Over the past ten years, we've all learned a great deal concerning sustainable supply chains. Certainly, today's corporate leaders and managers understand the business risks attendant upon sourcing commodities from the developing world. At its most basic level, any commodity such as cocoa, for example, needs to contribute positively to their lives if we expect the many millions of farmers growing such important crops to continue to want to do so. Equally important, they need to be grown responsibly if we expect customers and consumers to continue to purchase the products made from these commodities." (Long 2008, p. 317)

This statement of the Hershey Company's former CSR manager, John C. Long, from a strategic paper that he wrote to motivate leaders from other companies to join forces to improve sustainability in the GCCC, is particularly telling. It provides a special insight into the company's interest to shape perceptions and attitudes of both producers and consumers—a strategic objective that has been described as the consent creation-side of CSR as a governance tool. Furthermore, in his paper, Long stresses the pressing sustainability challenges which would constitute an immediate risk for the long-term supply of cocoa and jeopardize the whole industry. In order to confront this risk, he calls for the establishment of effective strategic alliances among the industry. In addition, dialogues across all stakeholder groups of the value chain including donors and implementers and a dedicated engagement in closer relationships with civil society actors are described as important prerequisites for the achievement of sustainability in the sector—all measures that can be explained with the theoretical framework of "consent and control-governance".

These two sides of governance aspirations respond to consumer pressures and handle the possibility of producers' reorientation towards other income strategies, as well as assure a sustainable flow of the produce can be spotted by having a closer look to the evolution of CSR in the GCCC. Prior to the first international indignation about child labor and other ethical grievances in the chain, the industry was already concerned with the long-term sustainability of cocoa. The industry's fear of a production shortfall is strongly linked to the high production losses in Brazil from 1989 onwards which were caused by a severe outbreak of the cocoa disease Witches' Broom which impacted the global market supply strongly. Driven by these developments, the first International Workshop on Sustainable Cocoa Growing, known as Panama Conference, was organized among main interest groups, such as industry, environmental organizations, foundations, universities, and agricultural research centers (Shapiro and Rosenquist 2004, p. 456 ff.). Interestingly, no government representatives were present but five principles were established which later on would guide the first public-private partnership arrangements. The conviction that arose among participants was that an improved cocoa cultivation would require a holistic approach and multifaceted efforts such as training, technology transfer, credits, and improved genetic material (Shapiro and Rosenquist 2004, p. 456 ff.). One year after the Panama Conference, there was another important multi-stakeholder meeting organized by the World Bank. During this meeting, representatives from the cocoa and chocolate industry highlighted the need for a stable long-term supply of cocoa which would require the support of small-scale farmers and the stabilization of tropical farms (Shapiro and Rosenquist 2004, p. 459).

After that established industry goal to improve control on the production base, in 2001 with the child labor scandal and the linked critiques on ethical grievances in the GCCC, the need to keep the broad consent on the functioning of the chain also arose. The above-described consumer pressure for more fairness in the chain and the results of the Harkin and Engels protocol made it necessary to find corresponding measures. As a result, many philanthropic projects have been implemented at the community level. In his strategic paper, Long (2008) describes the process which the industry went through: While companies initially tended to be rather reactive and just replied with short term CSR projects to outside pressures as for instance inquiries related to sustainability issues from advocacy groups and consumers, there was a shift towards becoming more strategic and developing industry-wide agenda in the precompetitive field of sustainable supply chains (Long 2008).

A closer look at the dynamics allows the understanding of the merging process of the two governmental needs to one pro-active CSR approach which enabled the industry to respond efficiently to both pressures which came on the agenda almost simultaneously. Initially CSR projects were rather isolated and most of the CSR projects concerned child labor and community support with infrastructure such as the building of schools, clinics, and water and sanitation facilities. But in the course of the 2000, projects became more sophisticated and a sector-wide understanding of sustainability requirements emerged. As a result, many CSR projects are now part of multi-stakeholder programs in a context of value chain restructuring which mainly focuses on the production level of the chain and the professionalization of cocoa farmers.

The overall panacea for the improvement of sustainability in the GCCC became the professionalization of the farmers. While leading agronomists do not necessarily agree with the proclaimed risk of cocoa shortfall and highlight the steady production increase over the past decades (Ruf and Bourgeois 2014), the industry sees the main problem as the farmers' traditional cultivation and post-harvest management techniques and attests a considerable yield loss due to local inefficiencies. Moreover, on many small-scale farms in West Africa, both cocoa farmers and trees are aging while the younger generation is less interested in continuing their family's cocoa business. The companies are aware that due to small income and hard physical work, attractiveness of cocoa farming decreases as soon as alternative income options are in reach (Anyidoho et al. 2012). Therefore, the emphasis is on the modernization of cocoa farming which would make the business more attractive for younger rural dwellers while simultaneously boosting the quantity and quality of production. To achieve this goal, farmer training in Farmer Field Schools with Good Agricultural Practices (GAP) became the most used approach. The aim to improve small-scale farmers' productivity is not only oriented towards supply sustainability but also towards social and environmental sustainability. The prevailing assumption is that farmers' poverty can be reduced through their "professionalization" and the resulting increase in their production. Similarly, environmentally damaging farming practices are supposed to also be overcome through the application of the GAPs which is an important part of the modernization of the sector. Hence, the professionalization of farmers became the main response to the sustainability deficiencies in the GCCC.

Therefore, it can be argued, both pressures—consumer awareness of ethical issues and the need for an improved sustainability of supply—in the 2000s led to a rapid move of TNCs towards the implementation of more proactive CSR strategies in the GCCC. The industry's aspirations to improve supply sustainability were increasingly integrated to traditional CSR projects. Multi-stakeholder programs such as the West Africa Sustainable Tree Crops Programme (STCP) and the WCF's Cocoa Livelihoods Programme played key roles in the progress towards this direction. CSR in its initial form of philanthropic community development projects no longer plays a significant role. Rather, the main focus of present-day CSR is about sustainable supply chain management.

Looking over the past two decades, CSR served as the major catalyzer for any kind of TNCs' sustainability interventions at the local level, in other words the partial backwards integration in the chain. Genier et al. describe two main approaches of CSR in the Agrifood sector: one, certification of the application of standards and codes, and two, value chain interventions. According to them, the former would fulfil the functions of increasing control along the chain and demonstrate sustainability efforts to consumers, and the latter would seek to scale up production and to encourage producers to participate in a global value chain. Figure 4.6 illustrates the evolution of the different CSR approaches in the GCCC over the past two decades. Traceability, GAP training and rural service centers all fall under value chain interventions whereas 3rd party certification actually is a combination of traceability and GAP training advanced with more sophisticated measures of documentation and control. These interventions and measures of the sustainable production approach will be topic of chapter 5.3. on the main challenges to establish sustainability in Ghana's cocoa industry. Their differences, how they are connected to each other, and more details will be provided there.

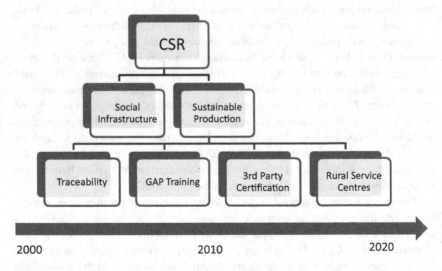

Figure 4.6   The evolution of CSR approaches in the GCCC. (Source: own elaboration based on expert interviews in 2015 and 2017)

Since the beginning of the 2010s, TNCs have proactively sought to shape the agricultural practices of the farmers and modernize the sector within the scope of their CSR strategies (cf. Barrientos 2011, p. 13). All major chocolate brands and main cocoa processors have implemented sophisticated CSR strategies which mostly combine aspects of community and social development with measures to achieve production increase and environmental protection. From the chocolate brand side, for example, there is Mars Incorporated's 'Sustainable Cocoa Initiative', Mondelez' 'Cocoa Life' or Nestlé's 'Cocoa Plan'. Equally, the grinders have their programs in place: For instance, Cargill runs its 'Cocoa Promise', Barry Callebaut the 'Cocoa Horizons' or Olam the 'Grow Cocoa Programme'. At the same time, consumers' demand for more equitable cocoa products has rapidly increased. Fair trade and sustainable production goods left their niche position and are now part of the mass consumption market. Already in 2012, 22% of the whole cocoa production was standard-compliant and 10% (Potts et al. 2014, p. 134) of the whole global export market was sold certified by one of the four main sustainability labels existing in the cocoa/chocolate market: Organic (0.2%) Fairtrade (4.5%), Rainforest Alliance (7.6%) or UTZ Certified (9,8%) of total cocoa production (Cosbey 2015, p. 12).

By the mid-2010s, most of the CSR projects applied certification programs. But in the field, a multitude of different approaches to manage sustainability systems are still present. Some companies work through NGOs, others in cooperation with government or donor agencies, while others enter into private-private partnership constellations or expand their inhouse programs. However, notwithstanding the different forms of managing the implementation, TNCs from the cocoa-chocolate industry apply similar approaches to their sustainability strategies. Among the most prevalent approaches are the following:

- Training of farmers with Good Agricultural Practices (GAP),
- the provision of improved planting material and support the access to other inputs,
- the introduction of new technologies as for instance mobile phones for the distribution of important farming information,
- the setting up and support of farmers-based organizations or the close cooperation with cooperatives,
- the application of third-party certification schemes, and

the cooperation with NGOs and industry stakeholders in a close partnership. Most of the above-mentioned CSR programs of the companies apply several of these approaches, but every company has a different emphasis.

In response to the TNCs' interventions in community-level socio-economic issues, an agricultural development analyst, cited by Barrientos, made this indicate statement: "Companies are now running around Africa like NGOs, but they don't really know what to do." (Barrientos 2011, p. 3)

Part II

CSR and Transformations in Ghana's Cocoa Sector as Part of Governance Dynamics in the Global Cocoa-Chocolate Chain

The Cocoa Sector in Ghana—a Frontier at its Limits?

The story of cocoa in Ghana is the story of a moving cocoa frontier which enabled unprecedented increase in production and catapulted Ghana to one of the world's top cocoa producing countries. After a strong cocoa recession in Ghana starting in the 1970 s and leading almost to the collapse of the industry in the country in the 1980 s, Ghana achieved its comeback and is today the second largest cocoa producer right after its neighboring country, Côte d'Ivoire. This second cocoa boom, however, was not only achieved through expansive migration movements into new virgin rainforest lands but also through a rapid and widespread adoption of intensification measures by the farmers. Ever since, Ghana's cocoa production is constantly growing.

Cocoa used to be Ghana's economic backbone for many decades in the 20th century and today it is still Ghana's major cash crop and one of the main export commodities after oil and gold (Central Intelligence Agency n.d.). About 800,000 farmers (World Bank 2013) cultivate cocoa on an average of 5 hectares (Hainmueller et al. 2011) and earn an average daily income of 0.84 USD (Fountain and Hütz-Adams 2015). Hence, even if cocoa production may to some extent have contributed to the reduction of poverty among rural dwellers (Kolavalli and Vigneri 2011, p. 206), the actual poverty rate among cocoa farmers is still high (Laven and Tyszler 2018) and subsequently, many cocoa farmers are still deprived from their basic needs and live in food insecurity (Dei Antwi et al. 2018; Knudsen 2007, p. 40). Next to cocoa farmers, about 30% (Hütz-Adams 2009) of the population in Ghana depend on activities linked to the cocoa sector, for instance in storage centers or in grading and transportation activities (Knudsen 2007, p. 35). Due to its constant delivery of high-quality beans and the reliability the Ghana Cocoa Marketing Board (COCOBOD) has achieved over years of its sales, Ghana holds a special position in the world market and receives an extra premium for

© The Author(s), under exclusive license to Springer Fachmedien Wiesbaden GmbH, part of Springer Nature 2023 119
F. Ollendorf, *The Transformative Potential of Corporate Social Responsibility in the Global Cocoa-Chocolate Chain*, (Re-)konstruktionen – Internationale und Globale Studien, https://doi.org/10.1007/978-3-658-43668-1_5

its high-quality beans (Kolavalli and Vigneri 2011; 2018). This is mainly the result of Ghana's response to the Structural Adjustment Programs in the 1980 s; a partial liberalization of its cocoa sector. Today, Ghana is one of the most flourishing economies in Sub-Saharan Africa and has also diversified its economy. Although cocoa makes only 2% of Ghana's GDP, it still plays an important role economically, contributing to 20 to 25% of the total export revenues (Kolavalli and Vigneri 2018, p. 4).

Being the world's second largest producer of cocoa beans and processed cocoa products, Ghana plays an important role for the transnational cocoa and chocolate industry. But due to complex challenges the sector faces, many stakeholders in the chain regard the future of cocoa production in Ghana to be at risk and see the need for constant and integrated efforts to maintain production rates at the level of those from the past decade (Wessel and Quist-Wessel 2015). Given the nature of a declining forest rent, that is the declining productivity of older cocoa farms, there are two main strategies to keep or raise productivity: the expansion of cocoa production into new forest areas or the intensification of production. The former option does not present as a feasible approach because the few remaining forest sites are now protected from deforestation through natural reserves and various regulations. Therefore, the focus of most of the industry stakeholders is on the intensification of already established farms. Ghana's second cocoa boom was mainly realized through intensification which rose to such an extent that leading scholars in the field saw the beginning of a Green Revolution in Ghana's cocoa sector (Ruf 2007). However, concerns about ecological and social side-effects of intensification have been raised.

The successful introduction of intensification measures in the early 2000 s goes back to two important developments in the Ghanaian cocoa sector: improvements in the efficiency of public sector programs from the Ghanaian Cocoa Marketing Board (Kolavalli and Vigneri 2018) and the rapid spread of TNCs' interventions in Ghana's rural areas, mainly in the course of their CSR and sustainability strategies. The present chapter seeks to give some information on these two dynamics and some historical developments which led to them. In order to better understand the nature of TNCs' concern to sustain a high level of production, the chapter first traces cocoa frontiers in Ghana and their role for rural transitions to more diversified economies. In a second step, the value chain in Ghana's cocoa sector, including its history and policy reforms, its institutional environment, and its participating stakeholders will be described as part of the value chain analysis. After a brief overview of the pressing sustainability challenges Ghana's cocoa sector is facing, the most prominent CSR approaches seeking to respond to them, and the general structure of CSR in the sector will then be described before finally turning to the empirical part of the study.

5.1   Ghana's Cocoa Frontier as a Driver of Rural Structural Change

Cocoa production in Ghana became the country's economic backbone for decades in the 20th century and was mainly driven by frontier movements. Being already cultivated on a large scale in the Americas since the early 19th century, cocoa was first introduced to West Africa by a group of Basel missionaries who distributed cocoa beans in Fernando Po. In 1876, the plantation worker Tetteh Quarshie brought cocoa seeds from the island to his village in the Eastern Region of the former Gold Coast which corresponds to the Geography of modern-day Ghana. Subsequently, cocoa quickly spread over the whole tropical belt of the former Gold Coast. Already in 1891, cocoa exports from the British colony had begun and production was as high as 40,000 tons per year (Ross 2014, p. 59), making the colony the world's largest producer by 1911 (Muojama 2016, p. 36). Production in the former Gold Coast exploded and in the 1930 s reached 300,000 tons per year (Austin 1996, p. 154).

Unlike in Latin American countries where cocoa was predominantly produced on huge plantations, cocoa in West Africa was cultivated by local inhabitants, mainly merchants and commercial farmers, on small and medium plots (Hill 1961). This rapid production increase was only possible because of large migration movements of farmers into the sparsely populated tropical forest belt, the process of establishment of agricultural pioneer frontiers (Knudsen and Agergaard 2015, p. 326). An agricultural frontier has been most broadly defined as "unexploited areas which are opened up for agricultural activities" (Barbier 2005). According to Argergaard et al. (2010, p. 3), frontiers are characterized by "rapid changes in demographic structure, economic basis, occupational possibilities, and land use" and attract people looking for new opportunities due their highly dynamic developments. A frontier crop is regarded to play a significant role in the development of the receiving rural area. Ruf (1995) developed the "cocoa cycles"-model. The model depicts the continuous shifting of production regions through the establishment of new cocoa frontiers in virgin forest areas, through which the several waves by which the cocoa production in Ghana spread until the final frontier in the far west of the country can be explained. The starting point of the model is the forest rent provided by highly fertile soils from tropical forests, which motivates migrants to move into sparsely populated tropical forest zones to establish new cocoa farms. The forest rent combined with the new labor force that enters the cocoa sector often leads to such a strong production increase that Ruf calls this phase a cocoa boom. The boom period may last for a few decades, however, both cocoa farms and farmers age and in combination with other ecological effects, such as the spread of cocoa diseases, yields generally decrease after three decades. The production decrease of a cocoa boom region has an effect on the global market and likely spurs higher prices for the more scarce produce. This, in turn, encourages new entrants to the sector. Hence, a new migration wave into new forest lands begins. This cycle from cocoa boom to cocoa recession and back to boom may occur within in the same country of production but also on a global scale in several countries.

Ghana has experienced such a cocoa boom which stemmed from the continuous migration of people willing to engage with cocoa into increasingly remote areas of the country. After Tetteh Quarshi established his cocoa farm in the Akwapim Ridge just north of Accra, the seeds spread throughout the Eastern Region (called the "old cocoa frontier", 1880 s). Merchants brought it to the powerful Ashanti kingdom and many chiefs began to participate in the cocoa business. Quickly, the Ashanti region became a dominant production area (1910 s) (Muojama 2016, p. 37). In the 1940 s, cocoa was also planted in the Central Region and Brong Ahafo region and finally, in the 1960 s, the first cocoa settlements were established in the Western Region, which is today Ghana's last cocoa frontier (and in 2017 was divided into Western North and South regions). In 1964/65, almost ten years after independence, Ghana produced 581,000 tons and was by far the world's largest cocoa producer. Ghana experienced one of the fasted expansions of a cash crop organized by local farmers that had ever occurred (cf. Ruf 1995, p. 177). But during the 1970 s, due to political crises in the country (cf. Mikell 1989), and strongly declining producer prices (Ruf 2007), production fell sharply and almost collapsed in the 1980 s (Kolavalli and Vigneri 2011, p. 201). In the 1980 s, severe droughts and bushfires in the Ashanti and Brong Ahafo regions expedited the production decline in these regions. This development accelerated the migration movements into the Western Region, which since 1984/1985 is Ghana's leading producer region (Ghana Cocoa Board 2020b). In the 1990 s, Ghana managed a comeback of its cocoa production that was pushed through a number of sector reforms. In the beginning of the 2000 s, cocoa production in Ghana was up to 600,000 tons, making it the second largest producer in the world again, right after its neighboring country Côte d'Ivoire. This immense production jump from around 160,000 tons annually in the early 1980 s (Kolavalli and Vigneri 2018, p. 2) can be regarded as Ghana's second cocoa boom (Ruf 2007). But this time, in contrast to the first cocoa boom in the first half of the 20th century which was solely based on farmers' migration movements and expansion of cocoa planted lands, the boom was achieved not only through such expansive frontier movements, but mainly through the introduction of intensification measures. According to Ruf (2007), the 50% production increase from 2001/2002 to 2006/2007 stemmed from about 15% increase in production areas and 35% in yield per hectare. The quick and widespread adoption of fertilizer and to a lesser extent insecticides and fungicides, among Ghanaian cocoa farmers and the adoption of some improved farming techniques were the main drivers of intensification. The expansion into the last available forest zones in the Western Region of the country brought new labor input to the sector and also the climate in the newly cultivated zones is very suitable bringing higher yields even during light crop season (Ruf 2007).

According to Knudsen (2007), high production districts such as the Juabeso district in the Western Region have increased their population by 50% with the cocoa migration flux. Cocoa is a smallholder activity but also gives opportunity for temporary labor, for instance for pruning, spraying, and harvesting activities. It therefore attracted both, migrant farmers, mainly from the older growing regions, who settled down, and farm workers who mainly came from the very poor Northern regions of Ghana and who had heard about the working opportunities.

Before the arrival of cocoa, the majority of the indigenous Sefwi population in the Western Region were subsistence farmers. Many of them then started to engage with cocoa too. There are complex dynamics of land acquisition linked to the process of cocoa migrant farmers' settlement. For migrants, it is difficult to obtain land from the local chiefs. Hence, most of them entered share tenancy relations and did not immediately become the owners of the land they farmed. Depending on the tenancy agreement, migrant farmers can keep revenues from half (Abunu system) or a third (Abusa system) of their harvest. Until today, migrant farmers are more economically vulnerable than indigenous farmers (Knudsen 2007).

The socio-economic and cultural life in a migration-receiving frontier is dynamic. Often, migrants settle together with other migrants in new villages and change the regional character. Negotiations about socio-economic arrangements between indigenous and migrant groups are crucial for the development process. Argergaard and Knudsen (2015) describe this process that is most characteristic for agricultural frontiers but still under-researched. The frontier gradually becomes delinked from the crop which initially triggered the frontier movement. This is what they observed to be currently happening in Ghana's last cocoa frontier, the Western Region, and what they identified to be linked to four migration flows. In the first phase, cocoa pioneers opened the frontier (1966–1979), followed by the second phase, a broader movement of cocoa farmers seeking to establish their own farms (1980–1989). Being motivated by the new working opportunities on cocoa farms, cocoa labor migrants also moved to the area, the third phase (1990–1999). Finally, from 2000 onwards, the frontier became so densely populated and lively that migrants looking for other business opportunities, especially trading and services were also increasingly attracted to the frontier (Knudsen and Agergaard 2015). The non-farming sector became more and more important and new opportunities for income diversification opened up.

Subsequently, the authors observe the emergence of rural agglomeration that brings many new livelihood opportunities for cocoa farmers and farm workers and keeps on attracting non-farm migrants. Once new opportunities exist and investment capital is available, cocoa farmers often tend to invest in off-farm businesses in order to diversify their income strategies (Knudsen and Fold 2011). The frontier region appears to be in a dynamic process of rural transition, triggered by the high influx of migrants motivated by cocoa. Until today, cocoa still plays an important role but every frontier forest rent will be exhausted after some time. If well applied, measures of intensification are able to outbalance the degrading fertility for a while. The declining forest rent in combination with the transition of the region into an economically more diverse area means that the future of cocoa is uncertain. The production shift into the Western Region was the last possible cocoa frontier shift in the country since remaining land is protected through forest reservations. In West Africa, the only remaining unexplored lands which would be suitable for cocoa lie in the very South West of Côte d'Ivoire and in Liberia (Ruf et al. 2015b). Hence, a production increase through the traditional cocoa cycles and the establishment of new cocoa frontiers is therefore almost impossible now. Given the rising land pressure, it appears that many stakeholders from the cocoa industry have a great interest in boosting the attractiveness of cocoa production and making it a business that young farmers would be eager to choose.

5.2   The Institutional Setting of the Cocoa Sector in Ghana

Looking at the history not only helps to understand the limits of cocoa frontiers in Ghana but also the functioning and institutional arrangement of the sector today. In this section, first a brief overview of the different institutional arrangements in the light of the political eras of the country will be given. Next, today's institutional set-up will be described paying a particular attention to stakeholder positions and responsibilities. This should provide the background to understand the functioning of the GCCC at the national and local levels in Ghana.

5.2.1 A Historical Snap of the Institutional Environment and Sector Reforms

Being such an important export crop since the colonial period, cocoa production was a major interest for all following regimes in Ghana and has been strongly shaped by the political eras in the country. Ton et al. (2008) carved out four main periods of distinct sector policy approaches: period 1 "colonial governance" (1920 to 1957), period 2 "state-controlled cocoa economy" (1957 to 1980), period 3 "gradual reforms" (1980 to 2000), and period 4 "shifts from within" (2000 to 2008).

During the early years of Ghana's first cocoa boom during colonization, the cocoa industry was a free-market system in which European companies controlled the local buying from farmers as well as the exporting of beans. But in the 1940 s, the British government increased its position in the sector governance by establishing several marketing boards. There seem to be two main reasons for the rapid increase of British government's control over the sector. The colonial Department of Agriculture initially encouraged the creation of cocoa farmers' cooperatives as it saw farmers' improved organization as an important tool for a good and steady supply of quality beans. But the farmers' cooperatives became powerful and in the 1930 s two general cocoa strikes were organized responding to sharp price declines. The British government set up a commission which was to report on the situation of the cocoa production in the Gold Coast and came to the recommendation that more colonial control was needed (Laven 2010, p. 79). At the same time, during World War II, the British government became concerned with the stability of cocoa sales due to weakened transport facilities and shrinking markets and began to intervene through own purchases (Kolavalli and Vigneri 2018, p. 1). In 1940, the West African Cocoa Control Board was established but replaced by the wider West African Produce Control Board in 1942 (Vellema et al. 2016, p. 236). In 1947, the Cocoa Marketing Board (COCOBOD) was created with the stated major aim of price stabilization. However, low prices were payed to cocoa farmers and large reserves with considerable economic and political impact built up for the colonial rulers (Kolavalli and Vigneri 2018, p. 2). Channeled through the COCOBOD, the British Ministry of Food became the only seller of cocoa and the buying companies, which were mainly the European companies, continued buying the cocoa.

The second period of the cocoa industry in Ghana is marked by the independence movement and post-independency decoupling from colonial control. However, after a few years, a sequence of military coups and regime shifts brought the overall economy and with it the cocoa sector into deep recession. Shortly prior to independence in 1957, the United Ghana Farmers' Council, which was seen as the "farmers' wing" of Nkrumah's Convention Peoples Party, was founded (Beckman 1976, p. 50). After gaining independence in 1961, the direct involvement of foreign buyers in cocoa purchases ended and the Farmers' Council had the monopoly on internal and external cocoa trade. The Nkrumah government went on to consolidate public efforts in the sector and many subsidiaries of COCOBOD were established during this period. The Cocoa Marketing Company (CMC) was set up which became responsible for the administration of cocoa sales and the registration of local buying agents (Amoah 1998, pp. 78–104). The government built a quality control system that enabled the country to receive a premium for the good quality of Ghanaian beans on the world market. Moreover, it established a pan-territorial fixed price policy which guaranteed producers a fixed price (Vellema et al. 2016, p. 236) but in times of high world market prices it would be lower than world market price. However, the government used a portion of the price gap to support farmers with input subsidies and services. It further invested in two national cocoa processing factories. With all these measures, the Ghanaian cocoa industry became completely independent from its colonizers. In 1966, the Nkrumah government was overthrown by a military coup and the Farmers' Council dissolved (Beckman 1976, pp. 11–17). It followed a long period of political instability and mismanagement from which the cocoa sector strongly suffered. Several changes had been undertaken by the respective regimes. First, the competition among local buyers was reintroduced but it did not last for long because of failures and delays with payments to farmers. In 1977, the single buying system was re-established, this time a subsidiary of COCOBOD, the Produce Buying Company (PBC), was in charge of all local purchases. As a heritage of the different policies over the regimes and periods, in the late 1970 s, a large number of COCOBOD subsidiaries were in place to support the sector and to provide services to farmers and employed a large number of people (Ton et al. 2008): the Cocoa Service Division (CSD), the Quality Control Division (QCD), the Cocoa Research Institution Ghana (CRIG), the Cocoa Marketing Company (CMC), and the Cocoa Producing Company (CPC) (Laven 2010, p. 80). However, the sector's economy was declining, and the above-mentioned droughts and bushfires exacerbated the already difficult situation. The overall national economy was close to collapse when Lieutenant Rawlings and his populist Provisional National Defense Council (PNDC) took power through a military coup on December 31th in 1981. Inheriting a national economy that had suffered from macroeconomic instability, strong currency overvaluation, and ineffective state interventions (Vigneri book, p. 34), Rawlings accepted the intervention of international donor organizations.

In 1983, Ghana's Economic Recovery Programme (ERP), containing the IMF and World Bank's policy conditionalities to grant Ghana financial assistance for the needed reform process, was introduced. Ghana's cocoa sector became an exception from the general pattern of Structural Adjustment Programs in the Global South. While becoming known as Africa's "star pupil" (Ofosu-Asare 2011, p. 65) of implementing structural adjustment measures, in contrast to the other major cocoa producing countries in West Africa, Côte d'Ivoire, Nigeria, and Cameroon, Ghana managed to take a strong negotiation position and to partially resist the donor community's pressures. Initially, the donor institutions demanded the complete dismantling of COCOBOD. But as experience from the other three cocoa producing countries had shown, after the radical abolishment of the boards and full liberalization of their cocoa sectors, the quality of the cocoa beans reduced drastically. Ghana was already known for its high-quality beans, receiving a quality premium from the global market. The need to keep the system of public quality control in place was one major argument that supported the position of restructuring instead of completely dismantling the existing public institutions (Ofosu-Asare 2011, p. 67). Thus, only gradual reforms in the sector instead of full liberalization were introduced, a process that made the sector known as the "partially liberalized cocoa sector in Ghana" (Kolavalli and Vigneri 2011). Through its Structural Adjustment Credit, the World Bank introduced a number of economic stabilization and structural adjustment requirements whereof two directly addressed the cocoa sector: i) the annual review of cocoa producer prices and (ii) the review of Ghana Cocoa Board's (COCOBOD) cocoa marketing cost (Ofosu-Asare 2011, p. 7). The main target for the cocoa sector was to increase the producers' share of export prices through the drastic reduction of COCOBOD's operational expenditures (Kolavalli and Vigneri 2018, p. 36). The three main measures to achieve this were

- An increase in producer price
- The reorganization of the cocoa marketing board
- A reduction of cocoa marketing board costs through privatization and the introduction of multiple buying systems (World Bank 1983)

COCOBOD was already fixing a pan-seasonal and pan-territorial producer price in advance of the harvest seasons based on forecasting of the revenues. With the reforms, this procedure was moved into a stakeholder process of price determination and the Producer Price Review Committee (PPRC) was established. Participants include relevant actors of the sector like representatives of farmers, Licensed Buying Companies, haulers, COCOBOD, and Ghana's Ministry of Finance under which COCOBOD acts. Importantly, processing and manufacturing companies are not included in the process (Fold 2002, p. 231; Kolavalli and Vigneri 2011, p. 5). The setting of the annual fixed producers' price is possible because of Ghana's position as a reliable supplier of high-quality beans and the willingness of international buyers to buy in advance based on a forward-selling system. The reorganization of COCOBOD and the aim to reduce its operational expenses was mainly sought to be achieved through divestment in state-owned cocoa industry facilities and the privatization of important services such as transport and input supply. Both strategies were followed during the 1980 s decade. For instance, COCOBOD sold the majority shares of its insecticide plant and merged its processing factory WAMCO with the German Hosta-Group which became majority shareholder (Ofosu-Asare 2011, p. 122). Furthermore, COCOBOD's staff level was drastically reduced from 120,000 in the early 1980 s to 5,500 in 2006 (International Monetary Fund 2009). COCOBOD's responsibility for the construction of roads in cocoa producing areas was transferred to the Ministry of Roads and Transportation as well as the supply of inputs to farmers handed over to the private sector (Kolavalli et al. 2012).

In a next sequence of reforms as part of a series of structural adjustment measures (Fold 2002, p. 231) further far-reaching changes were introduced to the cocoa sector from 1992 onwards. The major objective was to further reduce COCOBOD's expenditures and the role of its subsidiaries in the sector. In 1992, competition was reintroduced to the internal marketing system. Next to the publicly owned Produce Buying Company (PBC), Licensed Buying Companies (LBCs) were now allowed to purchase beans from the farmers, a system that is the main feature of what is called the partial liberalization of Ghana's cocoa sector. Even if there is competition on the domestic market, COCOBOD maintains its regulatory function by granting licenses to companies which seek to purchase cocoa beans locally. The LBCs have to pay the farmers the annual producer price, adhere to quality standards set by COCOBOD's subsidiary Quality Control Company (QCC), and sell the beans to the Cocoa Marketing Company (CMC), of which COCOBOD holds majority shares (Kolavalli et al. 2012, p. 13; Ton et al. 2008, p. 10). In 1999, another important reform was implemented. The extension services, formerly conducted by the Cocoa Service Division (CSD), were moved to the Ministry of Food and Agriculture as a compromise with the IMF and World Bank (Ofosu-Asare 2011, p. 123). Activities of CSD which were not explicitly extension services were reorganized in two new divisions of COCO-BOD, the Cocoa Swollen Shoot Virus Disease Control Unit (CSSVD), which assisted farmers with rehabilitation of virus infected farms, and the Seed Production Division (SPD) which until today is in charge of raising hybrid seedlings for replantation of farms (Laven 2010, p. 89; Kolavalli et al. 2012, p. 24).

After these gradual but far-reaching reforms, COCOBOD's responsibilities and subsidiaries were trimmed but its key responsibilities remained in place. COCO-BOD was still able to determine producers' prices and to keep a part of the Free on Board price (FOB) while increasing farmers' share of it.[1] The CMC continued to control and conduct external marketing and to coordinate the internal marketing. The final quality check remained the task of the Quality Control Company and some services were still provided to farmers (cf. World Bank 2011). In 1999, the Government of Ghana, then under Rawlings as a civilian president, approved the Cocoa Sector Development Strategy I (CSDS I), also known as Ghana's National Cocoa Plan. The strategy was developed by a multi-stakeholder task force consisting of government officials, farmers, cocoa buyers, and experts in areas such as finance, processing, and marketing. Four working groups were established which focused on i) production, research, and extension, ii) marketing, processing, and quality, iii) infrastructure and finance, and iv) taxation and pricing (Ghana Cocoa Board 2013). Main objectives agreed on were the increase of production to 700,000 tons annually, to raise the producer price to 70% of the FOB, to reduce export taxes, to deepen the competition in internal marketing, to allow the LBCs to export 30% of their purchases, to allow the local processors to import low grade and cheaper cocoa for processing, and to encourage processing through non-price incentives (Ghana Cocoa Board 2013). Rawlings' successor government under Kufuor continued the implementation of the CSDS I but did not allow the LBCs to participate in export activities. During the 2000 s, under Kufuor's administration, COCOBOD's activities once again expanded gradually. The two main programs that were launched were aimed at sustaining and developing cocoa productivity: the High Tech Fertilizer subsidiary program and the Cocoa Diseases and Pest Control Programme (CODAPEC) (Kolavalli et al. 2012, p. 24). Both programs were conducted by the CSSVD unit. The free mass-spraying of cocoa farms should improve the control of pests and diseases on farms and the free distribution of fertilizer should encourage farmers to apply fertilizers. Both measures, being criticized as being overly expensive, showed

---

[1] FOB is a price arrangement where the buying agent pays for transport costs.

positive results only after a few years (Ruf 2007). But a deficit of appropriate extension services became problematic. It was sought to improve efficiency of extension services through moving them to the Ministry of Food and Agriculture (MoFA). However, it turned out that this rather created a gap for cocoa farmers since MoFA's agents gave only general and not cocoa-specific advice (interview with COCOBOD staff 2015, for similar findings see Laven 2010, p. 89f.)

At the advanced state of the reforms, several institutional voids had emerged. This was particularly the case for the extension subsector which holds a key role for the professionalization of farmers and boosting productivity. Hence, many new actors, mainly international private actors and often in the form of partnerships, begun to enter the sector and fill the voids (Ton 2008, p. 10). For instance, private input providers began to train farmers on the application of their products. Also, the Sustainable Trees Crop Programme (STCP) was an important catalyzer for the spread of the first private extension services in Ghana's cocoa sector. The particular role CSR played in this regard will be treated in Section 5.4. In 2010, the CSDS I was reviewed, again in a multi-stakeholder process, and led to the development of the CSDS II. The general policy direction was kept in place. The national cocoa output target was raised to 1 million tons annually and the growth trend should be sustained. The need to establish a comprehensive inventory on cocoa resources, including farmers' details such as socioeconomic conditions, farm productivity, and farm size, was acknowledged with the strategy, welcoming the efforts of LBC's which mapped farms in their traceability programs. In CSDS II, the new issues that were recognized to have emerged over the 2000 s are traceability and certification, organic cocoa, as well as climate change (Ghana Cocoa Board 2014b). In order to improve their key function, in 2010, extension services were brought back to COCOBOD/CSSVD in a form of a public-private partnership arrangement. But by then, as a COCOBOD staff put it in an interview, the table was already full when COCOBOD entered the extension room again and it had to find a new seat which was not the driving seat (interview with COCOBOD staff in 2015).

5.2.2   Today's Institutional Set-up and Stakeholder Positions

The institutional structure and policy context remain largely the same after the revision of the Cocoa Sector Development Strategy. COCOBOD continues to play a strong managing role in the national cocoa chain which is combined with a critical role of the private sector for input supply, cocoa sales, and transportation.

COCOBOD is not only regulating these activities but also implementing programs of subsidized input distribution, conducting research, testing and approving important inputs such as fertilizer, spraying machines, and agro-chemicals, and providing funding for social interventions in and around cocoa communities (Ghana Cocoa Board 2014b). The main objective of the CSDS II is the enhancement of the country's cocoa production. The challenges for this task and the respective strategies to respond to them are subject of the Sections 6.3 and 6.4. In this subchapter, the institutional configuration, that is stakeholder positions and their responsibilities, will be described.

After restructuring, COCOBOD remains a complex and sophisticated institution. Its board of directors is appointed by the Government of Ghana. The head office consists of nine directorates and several departments. Furthermore, five subsidiaries and divisions run the operational activities of the board. They implement the main activities to reach the objectives of the board, that is the encouragement of cocoa production[2] and local processing, the conduct of research on cocoa quality as well as the approval of chemical inputs, the regulation of the internal marketing, the setting of policies to arrange purchase, grading, sealing, certification, sale and exports, as well as the marketing and exporting of cocoa beans and products (Ghana Cocoa Board 2020d). The operations are divided into the pre- and post-harvest sectors. In the field of pre-harvest, the Cocoa Research Institute of Ghana, the Seed Production Division, and the Cocoa Health and Extension Division are the responsible agents. For the field of post-harvest, the Quality Control Company and the Cocoa Marketing Company are in charge. The following short description of each entity[3] provides the basis for an understanding of the functioning of Ghana's cocoa sector and how CSR interventions currently transform it which is topic of the empirical part of this study.

Cocoa Research Institute of Ghana (CRIG)
Established in 1938, CRIG is the oldest agricultural research center in Africa and became a renowned center for the study of cocoa and other tree species. Its research includes all aspects of the industry from production over processing to the various utilizations of cocoa. Among the core competences are the research on technical innovations to improve yields and fine flavor of cocoa, the identification of new marketable consumer products, and by-products as well as the provision of recommendations on insect and pest management.

---

[2] COCOBOD is also in charge of coffee and shea nut production but the cocoa sector is by far the most important one.

[3] All information retrieved from Ghana Cocoa Board 2020c, last accessed on 12.06.2020.

Seed Production Division (SPD)

SPD is a result of the dissolution of the Cocoa Service Division in the course of the reforms. While extension services were moved to MoFA, the multiplication and distribution of improved cocoa planting materials was kept at the board. SPD currently runs 27 cocoa stations where hybrid cocoa seeds are produced. It has the mandate to raise 60 million cocoa seedlings for distribution to farmers free of charge.

Cocoa Health and Extension Division (CHED)

CHED was established three years after bringing extension back to COCOBOD in 2010. It is the successor of the Cocoa Swollen Shoot Virus Disease Control Unit (CSSVD) but its responsibilities are broader. Next to the survey and control of pests and diseases, it is also in charge of the rehabilitation of old and unproductive cocoa farms and the provision of extension services to farmers. The rehabilitation scheme involves activities such as assisting farmers to clear their farms of unproductive trees and replanting with hybrid varieties as well as teaching farmers the application of fertilizer and the adoption of modern agronomic practices and business approaches to cocoa farming. When bringing extension services back to COCOBOD, it was under the premise of a Public Private Partnership arrangement, named Cocoa Extension Public Private Partnership, CEPPP. CHED also holds the responsibility to manage this partnership program and to coordinate linked stakeholder activities.

Quality Control Company (QCC)

QCC is responsible for verifying the quality of cocoa and the adherence to quality standards. It conducts control measures at every stage of the supply chain and certifies Licensed Buying Companies' (LBCs) depots where cocoa is bulked and stored. At the depots, the first grading, sealing and certification will be done by QCC staff. In a next step, at the take-over points at the port warehouses (in Takoradi, Tema or the inland port Kaase), QCC staff does 'check-sampling' on which basis advice will be given to Cocoa Marketing Company whether the beans should be accepted from the Licensed Buying Company or not. Finally, another 'check-sampling' is conducted for all consignments before beans are finally exported.

Cocoa Marketing Company (CMC)

CMC's responsibility lies in the marketing and selling of cocoa beans; be it to local or international buyers. At the take-over points at the three ports, CMC purchases the beans from the LBCs and then it stocks them at the port warehouses until the shipment. It manages the exports and seeks to secure optimal prices and maximize foreign exchange revenues. It has its headquarters in Accra but also a satellite office in London.

Next to COCOBOD and its subsidiaries, farmers and Licensed Buying Companies (LBCs) are the most important stakeholders in the Ghanaian cocoa sector. However, there are also local and foreign input suppliers and a number of haulers active in the field.

Cocoa farmers
About 800,000 cocoa farmers build the basis of the Ghanaian cocoa industry. Their crucial role for the functioning of the whole industry has been highlighted in the upper sections of this study. The general picture of the Ghanaian cocoa farmer is that of an older male smallholder farmer with a lower educational level cultivating one or a few plots of two to three hectares applying traditional farming practices. While this picture holds in many cases, there is a higher diversity among cocoa farmers in Ghana that should be kept in mind. As mentioned already above, middle and large-size landholdings also exist, even if small farms constitute the vast majority of farms. For instance, Knudsen and Fold (2011) found landholdings up to 50 hectares. At the same time, there are different groups of cocoa farmers such as autochthone and different types of migrant farmers who all have different backgrounds in cocoa farming.

Women have always been highly involved in cocoa farming, but officially were often not recognized as they do not hold land titles. However, the situation is gradually changing, and more land is being transferred from husbands to their spouses (Barrientos 2016) or women can now inherit land. Even in Ghana's rural areas, partly due to improved transport and communication facilities and increasing rates of formal education among rural dwellers, traditional cultural practices become tangentially less important which reduces the gender gap. Women are increasingly holding lands on their own thus more female cocoa farmers are officially recognized and being targeted by programs.

For the same reason of improved transport and a higher rate of secondary schools in the rural areas, the new generation in cocoa was able to achieve a higher educational level as the former one. Anyidoho et al. (2012) point to the fact that the rural youth consist of different groups and depending on their socio-economic backgrounds, cocoa farming is either a saving strategy for the achievement of their occupational prospects outside the agricultural sector or a real livelihood strategy. In addition to rural dwellers either seeking to make their living through cocoa income, often in combination with diversified incomes from other sources, or using cocoa as a saving strategy, there are also a number of urban middle- and upper-class land owners who invest into cocoa as a capitalist strategy. These urban landowners mostly have caretakers on their land and only visit the farms once in a while to oversee the progress or when produce is being sold to the LBCs.[4]

Licensed Buying Companies (LBCs)
Since the liberalization of internal marketing, LBCs, approved by COCOBOD, hold the responsibility to purchase cocoa beans from the farm gates for the annually fixed producer price and to bring them to their local depots. At community level, LBCs operate through their Purchasing Clerks (PCs) who buy the beans from the farmers. The LBCs have to transport the beans to the three take-over centers either themselves or through the use of haulers. The activities of LBCs are strictly regulated. They are required to grade the beans they purchase according to their size and quality and to ensure that they have been properly dried after the fermentation process. The revenues they receive for selling the purchase to COCOBOD is based on a set amount per quantity. Due to this fixed margin, LBCs seek to maximize their profits through their logistic capabilities and purchasing competence (Fold 2008, p. 104) which allow for the maximization of volume (World Bank 2013, p. 12).

The operation of smaller local LBCs is made possible through the system of a subsidized loan which they can receive based on their market share. The loan allows LBCs with little own capital to finance their marketing activities before getting paid when handing over the beans to CMC. In the operational year 2015/ 2016 (Ghana Cocoa Board 2016), there were 46 registered LBCs out of which 40 were actively involved in the internal marketing of cocoa.

The Produce Buying Company (PBC), which was under COCOBOD until 1999 when it was transformed into a Public Limited Liability Company, since liberalization of internal marketing maintains the largest market share. In 2015/ 2016, with 30.89%, it was the leading buyer of cocoa beans in Ghana[5]. It was followed by Armajaro Ghana Limited, a subsidiary of Armajaro Asset management, a London-based commodity investment firm, which made 13.43% of the internal marketing, and Olam Ghana Limited, subsidiary of the Singapore-based food and agri-business company, accounting for 11.79%. It is important to see that even if there is high number of LBCs active in the internal marketing, a small

---

[4] In the interviews with representatives from international donor organizations engaged in the cocoa sector in Ghana, this strategy has been found to even being present among some staff members.

[5] No more recent annual report of COCOBOD is available at the time of writing.

number of LBCs are responsible for the purchase of over the half of the sector. Operations of 27 local LBCs are so marginal that together they account for only 3.98%. International LBCs are advantaged mainly due to their better logistics and easy and constant access to finance compared to small local LBCs which suffer from a substantial lack of finance (Fold and Larsen 2008, p. 105).

Haulers
Private transport companies evacuate the cocoa bags from the local LBC depots to one of the three takeover centers if LBCs decide to not to do it on their own. They too are payed based on a fixed price determined by the PPRC (World Bank 2013, p. 13).

Input suppliers
Even though COCOBOD distributes some subsidized inputs through its diverse programs to farmers, the biggest share of inputs that are applied is sold by the private sector. Most of the input suppliers are small-scale suppliers. They operate at the local level either through small shops or by visiting communities. Some suppliers work as licensed distributer for manufacturers or wholesale imports (World Bank 2013, p. 11).

Local processors
As described above, in the early 2000 s, transnational processors established grinding factories in Ghana and a smaller share is processed by Ghanaian companies. In 2015/2016, more than 200,000 tons were processed into semi-finished cocoa products in the country.[6] Cargill has developed the largest grinding capacity accounting for 28.05%, followed by Barry Callebaut with 27.94% and Olam with 15.10%. The Ghanaian processing company Niche could achieve a good share of 10.48% but the Ghanaian Cocoa Processing Company Limited only 1.73%. In Ghana, cocoa-processing companies have no direct commercial link with cocoa farmers since the LBCs conduct local purchases. Grinders have to buy all beans from CMC. However, there are ongoing negotiations to allow processors to import cheaper low-grade beans for local processing.

---

[6] All information about processing shares are retrieved from the COCOBOD annual report 2015/2016, Ghana Cocoa Board 2016.

5.2.3   The Two main Segments in the Sector: Production and Marketing

For their analysis of the organization of organic cocoa in Ghana, Glin et al. (2015) divide the sector in production and marketing segments. This division is useful for the present study and will be applied here. The following description of the production and marketing segments in the sector will summarize activities of all stakeholders in the conventional cocoa chain in Ghana.

The production segment in Ghana's cocoa sector
Next to the cultivation of cocoa on-farm which has been described in Section 4.2, the production segment is determined by all activities that facilitate and support farmers' efforts to produce a good quality and quantity of cocoa. Farmers can obtain improved planting material from the SPD-run cocoa station in their district. Fertilizers and plant protective chemicals are sold to farmers either by private input suppliers who in some cases teach their clients how to apply the products. In general, during the cultivation process, farmers are supposed to receive capacity-building and technical assistance from CHED extension agents (public EA) or private extension agents (private EA) who operate under CEPPP. The major approach of delivering extension services is through farmer group meetings where extension agents gather farmers in order to train them with some Good) Agricultural Practices (GAPs). These are known as Farmer Field Schools (FFS). In cooperation with the German Development Cooperation Agency, GIZ, since 2010 CHED runs a special approach called Farmers Business Schools, where a particular emphasis is given to strengthen cocoa farmers' business and farm management capacities. Furthermore, farmers are supposed to receive some spraying and fertilizer distributed by COCOBOD. In reality, not all services always reach all cocoa farmers and there are reports of corruption in the delivery of fertilizer bags and other input material. Furthermore, according to COCOBOD policies, farms in highly productive areas are to be served first in order to keep their high production up. There are some credit schemes in place but in general getting a credit is very difficult for cocoa farmers. After the ripe pods are harvested, post-harvest treatment begins what is crucial for the achievement of good quality beans. Compared to other cocoa producing countries, the fermentation process before the drying of the beans in Ghana lasts longer (up to six days).

The marketing segment in Ghana's cocoa sector
Farmers then bring their 63kg bags to the local Purchasing Clerk (PC). The number of LBCs ready to buy depends mainly of the size and selling strength of the community. Every LBC acts through their PC as a local buying arm. PCs, working on a commission basis for their respective LBC, are mostly people from the community with a slightly higher educational level. They were trained to conduct quality checks of the delivered beans and can either accept or reject farmers' produce. As Ansah et al. (2018) observed, limited in competition through prices, LBCs tend to compete for the farmers to sell to them often through non-economic incentives such as the provision of services or the offer of subsidized inputs or credits. Since it is the PC who is in direct contact with the farmers, he[7] is key in the mediation between farmers and the LBC. From the storage place at community level, the LBCs bring cocoa bags to their local depots where QCC enters for quality check and sealing. Subsequently, from the district level all cocoa beans are brought to the takeover centers where QCC conducts further quality checks before CMC buys the consignment from the LBC. Beans are finally sold to national (48,500 tons) and international buyers (the major five destinations are Netherland 167,419 tons, USA 58,692 tons, Malaysia 57,375 tons, Belgium 37,207 tons and UK 30,334 tons) (Ghana Cocoa Board 2020b).

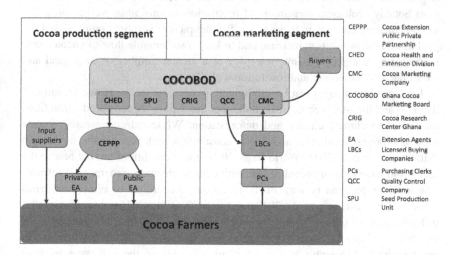

Figure 5.1   Major stakeholder relations in Ghana's cocoa sector. (Source: own elaboration)

---

[7] There are only very few female PCs in the sector.

Figure 5.1 shows a simplified representation of Ghana's cocoa sector, highlighting its two major segments, cocoa marketing, and cocoa production. COCOBOD is differentiated into its subsidiaries which fulfil distinct tasks in the sector. Not all direct links between the subsidiaries themselves and subsidiaries and farmers are represented but focus is given to the most important ones for the following analysis of CSR interventions in the sector. The dashed line gives indication that—due to the special position of COCOBOD—in the setting of the cocoa sector excluding CSR initiatives, international buyers did not have any direct link to cocoa farmers in both chain segments.

5.3   Main Challenges to Establish Sustainability of Ghana's Last Cocoa Frontier

While most of the industry stakeholders share the major common interests to keep a sustainable flow of high-quality beans or even to achieve a sustainable production increase (cf. Laven 2007) these goals depend on a great variety of factors which might role out even stronger in future. In Section 5.1, some aspects of the changing nature of Ghana's high production zone in the Western Region, also often referred to as Ghana's last cocoa frontier, were mentioned, most notably ecological constraints of production expansion as well as on-going socio-economic changes in these areas. But the picture is more complex, and the goal to achieve production increase and to keep a sustainable flow of cocoa beans is shaped by a highly dynamic interplay of a greater number of ecological and social but also economic and institutional challenges.

Ecological challenges can be subdivided into several facets. However, most of the mid- and long-term ecological challenges that Ghana's cocoa production faces are linked to climate change and deforestation. While extreme weather events such as droughts, bushfires, and floods constitute a high risk for the production (Monastyrnaya et al. 2016; World Bank 2013), these conditions are not new to the sector and have long impacted it. But with climate change a general shift towards new weather patterns is very likely to appear. According to climate scenario modelling by Läderach et al. (2013) and Schroth et al. (2016), there is a spatial differentiation of climate change impact within the West African cocoa belt. As has been outlined in Section 4.4.1, the most probable effect is an increase in temperature and droughts in some but not all areas of the belt whereby high temperatures are estimated to become more problematic for cocoa production than water availability. This is in line with what Ghanaian farmers report on their experiences with climate change (Ameyaw et al. 2018). Technical innovations such as hybrids and the needed chemical inputs are more efficient and bring higher returns in full-sun systems. They are hence more attractive for farmers who are in urgent need to improve their socio-economic position (Ruf 2011). But hybrid full-sun farms applying fertilizers are more vulnerable to climate change because of the missing shade canopy. Besides, these farms have a high production period of only 25 to 30 years. In addition, a high rate of chemical application comes with the risk of environmental degradation and pollution, consequences of intensification strategies. COCOBOD acknowledges the challenge of a declining soil quality (Oppong 2016). In most of the farms in the older production frontiers, soil fertility drastically declined after some decades and many cocoa farms are over-aged and need to be rehabilitated. But even after rehabilitation of these farms, the productivity rates have not returned to the high levels that fertile freshly cultivated forest soils offer. Other major biological challenges that constantly jeopardize cocoa farms are a number of severe pests and diseases. The two most important ones are black pod disease and the cocoa swollen shoot virus disease but capsids and weeds like mistletoes are also a risk for farms. In order to treat them well, additional labor input is required and/or investments in plant protection chemicals are needed.

In Section 4.4.2, the weak economical position of cocoa farmers was stressed. The poverty in which many cocoa farmers are trapped also shapes their ability to adopt improved farming methods and to invest in fertilizer, pesticides or seasonal workers. While many stakeholders and researches focus on low productivity as the major reason for extreme poverty among cocoa farmers, the impact that strong disparities in distribution of value within the GCCC have on farm gate prices gains little attention. In the present study, this link has already been described as a major cause of poverty among cocoa farmers in Section 4.4.2. Low farm gate prices lead to low household incomes and poverty which in turn are both regarded to be the main causes for poor working conditions and child labor on cocoa farms. Even if in Ghana COCOBOD takes the whole risk of price volatility on the global cocoa market, as it provides the farmers with a guaranteed price for the whole season and farmers do not have to handle temporary very low prices, the producer price is still too low to guarantee a fair living income. The Producer Price Review Committee determines the actual farmers' price of the season based on the FBO price which in turn is a result of the world market price. However, the outcome still does not enable cocoa farmers and their families to cover all their basic needs. A fair living income has to enable farmers to pay for decent housing, clothing, enough and healthy nutrition, safe drinking water and sanitation facilities, quality education, health care, transport costs and to have some savings for emergencies (Bahn and Schorling 2017). Hence, one of the core demands by the European Make Chocolate Fair campaign is the calculation of such a cocoa world market price which would at least provide fair living incomes for cocoa farmers.

The low income combined with the hard physical work is an important reason why cocoa farming is not an attractive business for many young rural dwellers in Ghana, an important social challenge for the future of cocoa production which is dependent on thousands of small-scale producers. Many elder cocoa farmers would not recommend that their children or grandchildren look for their future in cocoa farming (Hainmueller et al. 2011). The sector is known for having a high share of ageing farmers which are seen to be less efficient and less able to adopt new farming techniques. In a farmers' survey conducted by Anyidoho (2012), most of the younger rural dwellers included the survey stated that as soon as other income opportunities open up, they would abandon cocoa farming. Next to the little income, this also comes from their perception that the rural areas are boring places because of the infrastructure deficits. Even if medium-term development policies have led to improved rural infrastructure, especially regarding transportation and communication infrastructures (Schwartz 2013), in more remote areas, there is still a lack of basic infrastructures such as electricity, roads, education and health facilities, and mobile networks. These deficits have contributed to the high migration flux to urban centers in the cocoa regions resulting in labor shortages and high labor costs for cocoa farming (Dormon et al. 2004).

But Anyidoho (2012) points to the fact that the rural youth is a diverse group and several factors, such as the place of residence, level of education, whether migrant or autochthon or the travel experiences of the youth impact their occupational aspirations and life views. Particularly those young people with higher levels of education and with experiences in urban settings opt for formal work outside the agricultural sector while those young dwellers with some level of formal education would see farming as a means of capital accumulation needed for their non-farm occupational aspirations. According to Anyidohos's findings, full-time farming is rather the aspiration of the very poor youths who do not have any other perspectives. She further observed that, ironically, many agricultural training programs would rather target the more educated youth since they are regarded to better adopt new farming techniques. In general, the low level of education of many cocoa farmers puts them in a more vulnerable position, be it when buying necessary inputs or participating in Farmer Field and Business Schools.

A cocoa farmers baseline survey in 2011 (Hainmueller et al.) found that 33% of the survey households never attended any school and the average rate of years of schooling was 3.6 for males and 2.5 for females. Even if it is decreasing, the gender inequality that shapes women's positions in many Ghanaian rural societies is still an important social challenge in the cocoa sector. Traditionally, in many regions of the countries, land tenure was reserved for men so that until today, men constitute the majority of recognized farm owners (approximately only 20% of recognized cocoa farmers are female (Barrientos 2016, p. 3). On average, female-held plots are smaller and have less access to productive inputs (Vigneri and Holmes 2009). This is because the farm ownership is a prerequisite for many benefits such as receiving trainings and extension services or for the access to finance and the possession of farm passbooks which are needed to sell the produce for the LBCs (Barrientos 2016, p. 4). Hence, the gender imbalance in customary norms and practices translates into an even greater inequality in the currently institutionalized cocoa sector practices. This again brings many more constraints for women to increase their productivity and also makes them much more vulnerable to poverty.

Other challenges linked to the institutional setting of the cocoa sector include, for instance, the low degree of farmer organization and the limited access to input factors. The above-mentioned baseline survey found that only less than 10% of the survey population were organized in a farmers' association. This makes the reception of services such as trainings and the purchase of input factors such as chemicals and credits more complicated for them. The responsibility of input supply was handed over to the private sector under structural adjustment measures. This has led to some difficulties in supplying remote areas where the purchasing power is very low. In addition, there are many reports of fake products on the market. Especially farmers with very low educational levels and particularly those who are illiterate risk being cheated with very low-quality products. Purchasing inputs as a group would prevent farmers to step into such traps. Additionally, farmers' trainings would help to prevent such frauds and has already helped in the past. The special role that extension services play for improving farmers' knowledge and supporting the achievement of higher production has been acknowledged broadly (Baah et al. 2009; Onumah et al. 2013). Yet, at the end of the 2000 s, when the extension services were still allocated through the Ministry of Food and Agriculture, its reach to the cocoa sector was weak and malfunctioning. After moving back to COCOBOD and the reorganization of CSSVD into CHED in 2014, extension services have strongly improved but there are not enough direct visits to farmers by COCOBOD extension agents.

Further institutional challenges for the cocoa sector include illegal activities such as smuggling of cocoa beans to Côte d'Ivoire in the years when world market cocoa prices are higher than COCOBOD's annual producer price, and surface artisanal gold mining activities popularly referred to in Ghana as "galamsey" which are strong competitors for land (cf. Snapir et al. 2017). The existing policy gap of a missing land use plan has been emphasized in the World Bank's risk assessment report in 2013. Until now, the competition for land between different economic key sectors of the country is a pressing matter and, since land becomes increasingly scarce, will likely have a stronger effect on cocoa farming in the near future.

5.4   Main Sector Strategies to Respond to Sustainability Challenges

Stakeholders' strategies to respond to the above-mentioned environmental, economic, social, and institutional challenges are divers. While the Government of Ghana through COCOBOD and its various subsidiaries seeks to implement appropriate and efficient policies and strategies, many other stakeholders attempt to influence the sector in informal and less coordinated ways. Over the 2000 s, this led to quite a jumble and a lot of duplication of efforts at the local level of the sector. Subsequently, a number of institutions were established to improve sector coordination and cooperation between actors. This section gives and overview on existing strategies by the three main actor groups the public sector, civil society[8], and private actors.

5.4.1   Public Sector Strategies

In order to respond to the institutional challenge of missing coordination of stakeholder actions, three institutions have been established since 2013. The Ghana Cocoa Platform (GCP) is intended to be the main coordinative platform for sector exchange. Initiated in 2013 by the United Nations Development Programme (UNDP), the GCP was transferred to COCOBOD in 2015. The main aim of

---

[8] I decided to cluster farmers with civil society here even if they are also business actors. This is because I distinguish between social groupings' (farmers) strategies which seek to improve their living and working conditions from large companies' profit-seeking ones.

the platform is to enhance public-private stakeholder dialogue on the main barriers for scaling up sustainable cocoa production and joint-action planning. The main obstacles for sustainable production are defined as policy and institutional capacity for national technical support, the access to crop inputs, finance and markets, as well as the local land tenure systems (United Nations Development Programme 2013). Four corresponding technical committees were established: 1) Extension and Productivity Technical Committee, 2) Crop Financing, Marketing and Pricing, 3) Environmental and 4) Social Protection. The platform is financed by the UNDP, COCOBOD, the Dutch Sustainable Trade Initiative IDH, Mondelez International, and the World Cocoa Foundation (Ghana Cocoa Board 2020a).

Another institution within COCOBOD which seeks to foster public-private cooperation to respond to sector challenges is the Cocoa Extension Public-Private Partnership (CEPPP). CEPPP seeks to address the low rate of extension services to farmers. Formed with the return of extension to COCOBOD in 2011, CEPPP is under the responsibility of CHED. CEPPP aims to provide a framework for any private intervention at the local level[9]. Partners can either contribute funds to a common pool that is used for extension services and training of trainers, provide in kind support or capacity development or recruit and supervise their own staff to provide services to farmers under the supervision of CHED. All partner activities in the field of extension are supposed to be coordinated and monitored by COCOBOD. Private entities, mainly LBCs and international buyers (both processors and manufacturers), are invited to participate in the partnership but are not obliged to bring their sector efforts under the program. An interview with the program director in 2015 revealed that most of the companies do not use the CEPPP for their interventions but rather prefer to go on their own and without coordinating with COCOBOD. At the time of the interview, only Mondelez and Cargill as international buyers, the Ghanaian LBC Fedco, the German Development Cooperation GIZ, the Dutch NGO Solidaridad, and Rainforest Alliance were partnering with CEPPP. Important stakeholders in the sector such as the major LBCs Armajaro and Akuafo Adamfo or UTZ—the certification organization with the highest certification rates in Ghana- were not partners; showing the weak position of the program due to its voluntary approach. In reality, as was reported in the interview with the program director, the coordination and monitoring of partner interventions is a challenge, let alone doing so for activities from non-participating actors. Since most of the private actors are operating

---

[9] This information is obtained from the non-published "Guidelines for the preparation and implementation of Memorandum of Understanding for the Cocoa Extension Public Private Partnership" document.

independently from CEPPP, COCOBOD established the Programme Coordination Unit (PCU) in 2014 in order to gather information on and gain better regulation capacities of the many private interventions in the field.

Strategies to respond to environmental challenges coordinated by COCOBOD are implemented through a number of public sector programs. For instance, given that around 25% of all cocoa farms are over their economic life span and about 17% are infected with CSSVD, one major program is the cocoa rehabilitation program. Furthermore, COCOBOD continues with the CODAPEC program and the supply of chemical and organic fertilizers in order to assist farmers to handle pests and diseases as well as declining soil fertility (Oppong 2016).

When it comes to social sector challenges, COCOBOD is mainly concerned with social demographics of cocoa farms such as the high mean age of cocoa farmers and the lack of interest that many rural youths have in cocoa farming. Consequently, COCOBOD launched the Youth in Cocoa Farming Initiative in 2014. The initiative mainly supplies participating youths with improved seedlings, crop inputs, and technical advice to encourage their cocoa farming. Seeking to reduce rural-urban migration, measures of community development, such as improvement of roads networks and provision of schools, are part of the efforts (Oppong 2016).

By the time of writing, no distinct strategies to improve the economic situation of the farmers are spelled out by COCOBOD what suggests the view that economic challenges are a result of deficits in production (environmental and social challenges) and institutional sides (mainly the low extension to farmer ratio).

5.4.2   Farmers' Self-organization and NGOs' Support

Farmers' self-organization has been recognized as an important attribute of the resilience of a farming system which stems the basis for a sustainable global food value chain. Self-organization facilitates the coordination and expression of farmers' needs and interests improving their bargaining position in the sector and finally in the value chain. Furthermore, self-organization eases the spread of information and knowledge among farmers as well as between farmers and other stakeholder groups. From a practical point of view, it also simplifies the distribution of key services and inputs to farmers such as extension, chemicals, equipment as well as financial support.

In Ghana, cocoa farmers' formal organization exists to a very small degree (Laven 2010, p. 91). Historical circumstances have contributed to a generally skeptical perspective of cocoa farmers on institutionalized cooperation such as in the form of cooperatives or associations. In contrast to the neighboring country Côte d'Ivoire, where farmers' organization in formal cooperatives is widespread, in Ghana it has played a marginal role since the dissolution of the United Ghana Farmers' Council when Nkrumah was overthrown. However, there are currently three bigger cocoa associations and cooperatives and a few smaller ones mainly linked to the purchase of organic cocoa. The oldest and local one is the Ghana Cocoa, Coffee, and Shea-Nut Farmers Association, established in 1980. The other two important cooperatives are Kuapa Kokoo Farmer Union and Abrabopa. Kuapa Kokoo was founded in 1993 after liberalization of local marketing with the support of international NGOs such as the Dutch SNV, the British "development through trade NGO" Twin, and the British charity organization Christian Aid. Today Kuapa Kokoo has over 100,000 members and its commercial arm, Kuapa Kokoo Limited, is one of the major LBCs in Ghana (Kuapa Kokoo 2017). Abrabopa was established in 2008 by the Dutch-Ghanaian agricultural company Wienco and currently has about 7,000 members.

In Ghana, another l form of farmers' cooperation has evolved over the years. Cocoa farmers frequently formed groups around the cocoa buying centers of the Produce Buying Company (PBC) which became known as "societies". These societies have a clear organizational structure and elect committees consisting of chairman, treasurer, secretary, and four other members (Baah 2008, p. 397). With the liberalization of internal marketing, smaller societies also were established around other LBCs but to a lesser extent. By the end of the 2010 s decade, with the spread of certification programs, many LBCs engaged in such programs have engaged in setting up farmers societies for the implementation of certification activities.

Farmers' support from international NGOs has a long history and became widespread in the 1970 s and 1980 s when many humanitarian relief and development NGOs were founded in the Global North. In Ghana's cocoa sector such big international NGOs as the US-based religious World Vision or the Swiss-based charity organization Care International were primarily active in the fields of community development. The range of NGOs has broadened substantially and there is a diversity of NGOs which respectively follow different approaches and objectives. In Ghana's cocoa sector, one can observe the presence of at least three different types of international NGOs which sometimes cooperate with local smaller NGOs: the "traditional" humanitarian ones, industry-driven ones, and certification organizations which are perceived by many locals as NGOs and formally fall in this rubric but will be treated separately in the present study. The provided division does not seek to be exhaustive but rather presents what local interview partners perceived and what has been captured by the PCU team with its data collection. It is important to assess what type of NGO is implementing what kind of project if the objective is to understand power dynamics in the chain. Very often, the term NGOs gives a face of civil society representation to an organization which actually promotes industry values and market-based approaches without a prior civil society debate on developmental concepts.

5.4.3   Industry Interventions: CSR as a Response to Challenges on Ghana's Last Cocoa Frontier

As highlighted in several parts above and captured by the hypotheses guiding the present study, CSR is regarded here as an important tool for TNCs from the GCCC to respond to most pressing challenges and to ensure consent and control which is needed to sustain the flow of enough high-quality cocoa beans. In the case of the particular setting of Ghana's cocoa sector, CSR enabled TNCs to reach the local level and to respond to most pressing challenges in the last cocoa frontier in the country.[10] Since its rise in the early 2000 s, CSR in cocoa has undergone a shift from initially isolated projects of community development to sophisticated sustainability strategies covering a panel of interventions which are professionally tailored in TNCs' CSR or sustainability departments. These interventions seek to improve social and environmental conditions and to close the institutional gaps (see Section 4.5). Economic challenges, that is mainly farmers' poverty, are generally regarded to be a result of low productivity and is supposed to be overcome through farmers' efforts to achieve production increases. In the course of the focus shift and the sophistication of programs, CSR interventions have created a new institutional environment next to, and sometimes intertwined with, the already existing one. How CSR actually transformed the institutional setting at the local level in Ghana will be shown with the empirical part of the study and discussed subsequently. At this stage, only a broader overview of the different approaches will be given.

---

[10] The biggest share of CSR interventions in the cocoa sector is located in the Western and Western North Regions of the country.

COCOBOD's Project Coordination Unit (PCU) conducted a first assessment of the state of art of private (mainly foreign) interventions in the cocoa sector in 2014. The result does not capture all existing activities that went on in field by then but should come close to it. PCU staff members travelled the country for several months in order to obtain the information. In interviews with them, they reported the reluctance of many companies to provide the respective data on their interventions when contacted via phone or email. Altogether, PCU found 20 actors which were implementing a total of 52 projects, some of them large ones operating in almost all cocoa districts, some rather focused on specific areas, but with most interventions being implemented in the former Western Region (still containing the parts of Western North, too). Actor groups could be clustered into LBCs (5), NGOs (5), TNCs (5), farmer associations (3), and industry umbrella organization (2). The program types that were differentiated by PCU are input supply, forest conservation, certification, sustainability, extension, productivity, livelihoods, literacy promotion, child labor, organic cocoa certification, and traceability.

Based on the report, a pattern becomes visible: all actors work in partnerships. When it comes to activities directed towards productivity and farming inputs, most foreign actors work through partners who have a local link to the farmers. These are mainly farmer associations (from here onwards farmer-based organizations, FBO, which includes all different types of farmers' organizations) or LBCs. When it comes to activities of community development, NGOs are the implementers most often contracted by companies. The three certification organizations, UTZ, Rainforest Alliance, and fairtrade, are present too but do not implement the projects on their own. Rather, they partner with the companies which seek to purchase certified beans from farmers targeted by their projects. Some actors work closely with COCOBOD, some loosely and some without cooperation with the public entity. In the frame of the interventions, partnerships with different subsidiaries of COCOBOD are established. When it comes to traceability, cooperation with CMC is needed, when it comes to input supply cooperation with CRIG and CHED might be established. The picture of cooperation is diverse and complex. The following figure shows the main detected relationships linked to CSR interventions. In the empirical part of the study in Chapter 7, one UTZ sustainability certification project will be analyzed and evolving structures for the case carved out.

Figure 5.2 shows that TNCs implement their CSR strategies mainly through their local partners, that is LBCs, NGOs, or FBOs. In some cases, they seek cooperation with COCOBOD's subsidiaries for particular project aspects. Some

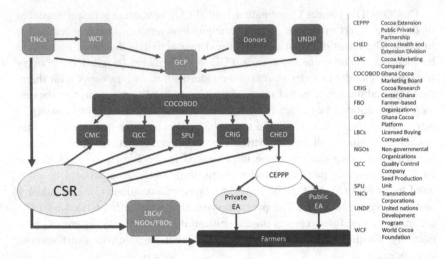

Figure 5.2 CSR links and stakeholder relations in Ghana's cocoa sector. (Source: own elaboration)

companies pool resources to CHED's Community Extension Public Private Partnership Program (CEPPP) and finance or send private agents for Extension and Advisory Service Provision (EASP). International donors and WCF cooperate with COCOBOD headquarters and on the Ghana Cocoa Platform for policy development and sector-wide strategies. WCF, besides advocating for business interests at COCOBOD platforms, also channels funds from the Bill and Melinda Gates Foundation or USAID on a matching grant mechanism to companies which will then be responsible for the implementation of respective programs.

In the following, four major types of existing different CSR approaches will be presented which summarize the types identified by PCU combined with information from own field work in 2015 and 2017. Due to the uncoordinated and large number of CSR interventions in Ghana's cocoa sector, the data basis is rather fragile and little literature has been identified which could be used additionally to the PCU and own findings. Hence, the following is an attempt to cluster transnational CSR interventions in Ghana's cocoa sector and does not claim to be exhaustive.

CSR for community development
The initial form of CSR in Ghana's cocoa sector was motivated by the need to show action against child labor. Projects of community development and the provision of important infrastructure such as school buildings, health centers, or water and sanitation facilities were built in cocoa producing communities to which no marketing link existed. Over the years, community development became less "hardware oriented" and more projects seeking to empower communities and enhance cocoa farmers' education are being implemented. For instance, child labor is often thought to be reduced by a general livelihood improvement and the empowerment of women who are seen to be key for communities' flourishing. Projects of literacy promotion and support of food crops and income diversification are typical examples of such projects. Often, these interventions are framed as the promotion of human rights.

CSR for production increase
Some companies, like Lindt and Sprüngli, conduct farmer training projects mainly in cooperation with LBCs (by the time of writing Amarjaro/Ecom) without using a certification scheme. The projects may include other aspects of service delivery such as the support to access farm inputs or credit. Often, the farmer training initiatives will be combined with a traceability approach which allows the company to follow the beans it invested in. If interventions are to put in a timeline, this approach probably was more applied before the "certification boom," or before 2010.

CSR through third party sustainability certification
Since the 2010 s, due to their commitments to source up to 100% sustainably produced cocoa, all the major manufacturers and processors apply one or more certification schemes to pursue their CSR strategies. Next to proving consumers the sustainability efforts through the effectualness of a seal, certification provides efficient responses to many of the described challenges. Generally, the procedure of becoming certified can be either initiated by individual farmers (in case of large-scale farming) or farmers' groups or by other value chain stakeholders. In the case of smallholder cocoa production, the idea of certification is mainly brought to farmers from downstream actors and rarely from their own initiative. Depending of which certification system is applied, the emphasis differs slightly (as described in Section 4.3). However, the main aspects include training programs with Good Agricultural Practices, Good Social Practices, and Good Environmental Practices (cf. Dohmen et al. 2014). The core steps of the processes

a TNC has to go through in order to get a certified produce from a farmers group are the following:

- Awareness creation and organization of a farmers society,
- setting up an efficient internal management system,
- training of farmers and getting the members applying the critical requirements,
- organizing an internal inspection of participants' farms, and
- organizing audit by independent inspectors to receive the certificate (Dohmen et al. 2014, p. 22).

A detailed analysis of this process follows in Chapter 7 on the case study.

Integrated CSR approaches

Integrated approaches combine productivity objectives with measures of community development and environmental protection. Such holistic programs are mostly realized in broader partnerships. An important role here used to play the WCF which runs, among other programs, the Cocoa Livelihood Program and the African Cocoa Initiative. As their names indicate, the former was initially more concerned with community development measures such as promotion of food crops and the empowerment of women, the latter put more emphasis on production increase through quality planting material and the provision of financial services. But in their advanced states (CLP 1 started in 2010 and ACI 1 in 2013, they are both in their second rounds now), both programs have incorporated aspects of the other side such as the provision of a 'full-package' of services to farmers through CLP and a strong focus on women and youth farmers in ACI (World Cocoa Foundation n.d.b). But over the years, individual companies have expanded their strategies to such an extent that they now cover most of the aspects linked to the above presented sustainability challenges. Programs applying highly integrated approaches include the Mondelez Cocoa Life or the Cargill Cocoa Promise.

Field Work Design and Procedures

6

The overall research interest of the present study is to estimate if CSR serves TNCs as a tool of private sustainability governance that seeks to increase their control over a given local production context and foster the consent for this intention among stakeholders. Therefore, the study seeks to reveal the mechanisms through which CSR transforms the local institutional setting and to understand how this process shapes the future perspectives of cocoa farmers. Inspired by Global Justice Theory, which highlights the importance to empirically assess perspectives on global institutions of those affected by them, it is a core aim of the present study to give room to CSR-targeted and non-targeted farmers and other sector stakeholders to articulate their concerns about and experiences with CSR. Therefore, the role the empirical data plays in the present study is twofold. First, the results of the data analysis seek to feed into the theoretical discussion on CSR as hegemonic governmental tool. This discussion will strongly benefit from empirical insights since there is almost no existing knowledge of these soft governmental strategies and their local outcomes available. Second, the documentation of some individual experiences, which are able to show to the distant chocolate consumer the problems on the ground, is also an important aim of the study. The importance of the empirical part for the study is balanced with the theoretical underpinnings on CSR as a governance strategy. It is the liaison of the two parts that enables the conduction of an informed and careful discussion at the end of the study. This chapter presents the methodical framework, the research procedures, the design of the empirical part of the study, including its guiding framework, as well as the characteristics of the study area and interview partners.

Supplementary Information The online version contains supplementary material available at https://doi.org/10.1007/978-3-658-43668-1_6.

6.1   Research Approach

Qualitative research approaches a particular social situation, event or process from its inner perspective. The main intent is to understand the lifeworld of acting people from their own views and experiences (cf. Flick et al. 2019, p. 14). As Creswell puts it: "The focus of qualitative research is on participant's perceptions and experiences, and the way they make sense of their lives. The attempt is therefore to understand not one, but multiple realities." (Creswell 1994, p. 162)

During this investigative process, typical qualitative research tools are open narrative or guided interviews, group discussions, ethnography, (non-) participatory observations, documentation of interactions etc. (Flick et al. 2019, p. 19). In contrast to a questionnaire, where research items are fixed and no additional information can be gathered, these instruments mainly aim to allow for as much openness as possible during the research process. It is this openness for the new and unknown, or as Meinefeld (2019, 266) puts it, the openness for the principles of construction of the experienced lifeworld, that makes qualitative research so important for many fields where theoretical knowledge seems to be too far from the individual contexts and where no solid information base is available for the researcher.

While quantitative research needs prior knowledge on the research object, qualitative research does not require the same. In fact, in the past decades, researchers encouraged that no prior knowledge should be included to the qualitative research process at all. However, today, this key dogmatic distinction of qualitative methods from quantitative has become less important. It was partly the result of the struggle of qualitative researchers for the recognition of the qualitative approach as a valid research methodology. However, nowadays, the standpoint that the openness for the new is not dependent on the researcher's prior knowledge but from the way and manner in which the search for the new is methodically designed, is readily more accepted. Within the several qualitative approaches, there is a stream that allows the researcher to predetermine the focus of their empirical study and to specialize the data collection on distinct research interest related aspects. In such a methodical setting, ex-ante hypotheses are seen to be of benefit for the research process (Meinefeld 2019, p. 273).

The present study follows such an understanding of qualitative research and applies the methodical set for the conduct of semi-structured key informant interviews and qualitative content analysis developed by the German social scientists Jochen Gläser and Grit Laudel (2012). According to them, qualitative research must not miss the benefits of setting up a hypothetical model and formulating research-guiding hypotheses. In that case, the researcher uses existing theoretical knowledge to gain a more precise idea of the empirical information needed to answer the research questions. The theoretical background is sharpened and accumulated in a hypothetical model on the assumed causal mechanism. Yet, this hypothetical model is not to be tested in the field. It only provides guidance and should direct the researchers' attention to the empirical circumstances while asking their research questions (Gläser and Laudel 2012, p. 77). A qualitative research design using such a guiding framework, which is not to be falsified, is still marked by its openness in the field. The set of variables, dimensions, and indicators is not static and can be readjusted anytime during the research process. Ideally, the research process is conducted during several phases, where the researcher commutes between field and desk works and stepwise develops the deepened understanding of the research object. Applying a stepwise approach but being guided by a theory-based causal model, the study is located between the logics of deductive and inductive approaches to the field. The study is not blind to already existing knowledge on the research topic and at the same times is still open for the inclusion of new items in the concept in the course of the research process.

The specific research objectives which arose from an ongoing theoretical and practical discussion on TNCs' societal responsibilities and the epistemological interest of the present study make the application of ex-ante hypotheses plausible. While the above-developed theoretical consent and control framework is not to be verified in the field, it provides a fruitful background to guide the search for specific mechanisms and sources of local transformations in Ghana's cocoa sector. The research questions require a mechanism-oriented rather than a relation-oriented explanatory strategy. The main interest lies in the understanding of the process through which CSR works locally as a governmental tool. Consequently, an understanding of which mechanisms are applied during implementation and to which structural transformation they lead is required. Such a study that seeks to reveal underlying mechanisms of causal relations generally requires *qualitative* research tools in order to grasp the *qualities* of such processes.

6.2   Guiding Model for the Empirical Assessment of CSR-induced Transformations in the Global Cocoa-Chocolate Chain

The elaborated conceptual framework on governance through CSR laid the basis for a more nuanced understanding of private governance in global value chains. For the present empirical study of CSR-induced transformations of the GCCC at the local level in Ghana, the amalgamation of Global Value Chain Analysis and neo-Gramscian perspectives on global governance to the consent and control framework resulted in a guiding model which digests the eight axes outlined in Section 3.3 in its different components (variables).

According to Gläser and Laudel (2012), the mechanism-oriented causal model seeks to reveal important social actions which mediate between cause and outcome, that is independent and dependent variables. It is based on four different types of variables, namely the independent variable which is the variable that provokes an effect, the dependent variable which is affected, the intervening variables which surround the scene and influence the causal mechanism, and the variables which describe the social actions which mediate between cause and effect (Gläser and Laudel 2012, p. 81). Figure 6.1 visualizes the scheme with its application to the context of CSR in Ghana's cocoa sector, the operationalization of the consent and control framework as well as the variables are explained subsequently.

Getting back to the theoretical underpinnings presented in Chapter 3, the main feature that the guiding causal model represents is the understanding of hegemony as a transformative social process that seeks to increase consent and control in a given value chain. To capture the ambitions of the present study and best operationalize the empirical section, it is necessary to translate a central question of neo-Gramscian theory to the context of the present study: how does a hegemony-aspiring group manage to generalize its particular interests and gain consent for the hegemonic status quo? This, in its end, is similar to the two main research questions of the present study (cf. introductory chapter):

1) *Does CSR serve TNCs as a tool to expand their power position in a value chain and if yes, how does it do so?*
2) *How and by which means is the implementation of a respective CSR intervention (certification scheme) achieved and how does this new arrangement transform the institutional setting of the Global Cocoa-Chocolate Chain at the local level in Ghana?*

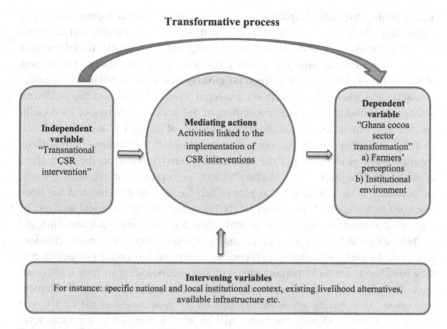

Figure 6.1 Guiding model on CSR-induced cocoa sector transformation in Ghana. (Source: own elaboration)

3) *How does the CSR strategy affect the farmers' perceptions towards their liveli-hoods and their cocoa production, and does cocoa production become more attractive to them again?*

Hence, how do TNCs from the GCCC use CSR to improve their sustainabil-ity governance of the chain with particular emphasis on the local level? The generalization of particular interests can be interpreted as a process of improving governance and making value chain stakeholders follow TNCs' goals of improved "chain sustainability". In the context of transnational CSR, the notion of sus-tainability itself appears as neutral but is generally defined by TNCs and allied stakeholders, therefore constituting a **topic of consent creation**. TNCs from the GCCC are understood as actors that seek to increase control along their value chain but particularly over locally dispersed smallholder production in a frontier at risk. Furthermore, the questions how and by which means CSR is imple-mented at the local level are addressed, thus, putting emphasis on the specific

action fields. Particular emphasis is given to the production **inputs** which are delivered through linked to CSR activities and to changes regarding **output** quality and quantity. Attention is paid to the way the local **institutional context** shapes the design and implementation process of the interventions. The **means of action** imply a number of **industrial governance** aspects, such as standards, contracts, and traceability which are analyzed in detail based on the fieldwork findings. When assessing the improvement of value chain governance as done in Chapter 8, the main focus is on the transformation of the GCCC at the local level. Thereby, two main aspects receive particular attention: the farmers' **perceptions** towards cocoa production and the **institutional environment** in the rural areas of the sector. The importance of stakeholders' perceptions and attitudes towards the dominant system is an integral part of neo-Gramscian thinking and has been described in Section 3.2. In Winter's (2007) concept for an empirical analysis of hegemonic movements, the notion of perception does not gain much attention and is likely subsumed in the axes of means of action or topic of consent creation. However, the shifts of farmers' perceptions with regard to cocoa production and their livelihoods are an important aspect of the understanding of how CSR functions. Therefore, the consent and control framework is extended with the element perception. After having given background on the functioning of the GCCC and the cocoa sector in Ghana, the focus will be narrowed down to the local level of production in order to reveal social mechanisms of influence and change. The topic of **scales** is thus progressively treated in the present study by highlighting and describing their interconnectivity in detail but with a distinct focus on the local level scale. The **actors** involved, their interests, the relations they have with one another, the new forms of cooperation established in the course of the CSR intervention, and so on are all crucial aspects of the following analysis (Table 6.1).

The guiding model is only a guidance during the data collection and analysis. It is important to highlight that there is no intention of testing it. Dimensions and components of the variables helped to design the interview guidelines and to direct the data analysis to remain close to the research interest. Dimensions and components were developed based on the consent and control-axes as well as information obtained during desk research, which has been presented in the previous Chapters 4 and 5. In the following, definitions of the variables and an overview of their dimensions and components will be provided. Interfering variables are not defined theoretically but their existence and characteristics are considered during the whole research process. Most influential interfering variables, such as the specific national and local institutional context, existing livelihood alternatives, and available infrastructure have been described in detail in Chapter 5. They are taken into account along the whole research process.

Table 6.1 The extended consent and control framework

Analytical elements of governance through consent and control							
Actors	Geographical scope	Industrial governance	Input-output structure	Institutional environment	Means of action	Perceptions	Scales	Topic of consent creation

Source: own elaboration.

Independent variable "Transnational CSR intervention"
The understanding of the mechanisms of social action by which CSR interventions in the GCCC are implemented and how these transform the Ghanaian cocoa sector is the main research interest of this study. Thus, in our causal model, CSR becomes the independent variable which mechanisms of operation and effects are the core interest. In the theoretical chapters, CSR strategies have been described as TNCs' tools for value chain governance. Accordingly, the working definition of the independent variable "Transnational CSR intervention" is as follows:

Transnational CSR interventions in the Global Cocoa-Chocolate Chain are strategic activities of TNCs which seek to improve the governance of sustainability challenges, that is to enhance their field of influence in order to increase control over local cocoa production and to establish consent on the status quo in the chain.

The difference in this definition to that of the common CSR definitions presented in Chapter 2 is quite sharp, but it is the focus of the study to explore the extent to which this side of CSR exists. As has been discussed in detail above, in the given context of the study, CSR is neither understood as a philanthropic action nor as a mere sustainability intervention. While it does aim to diminish poverty among cocoa farmers and to improve the ecological footprint of cocoa production these objectives are interpreted here as part of consent and control aspirations of TNCs rather than intrinsic goals. By reducing social and environmental pressures, CSR may serve as a means to foster consent for the status quo in the GCCC and simultaneously improve general control over local cocoa production and production flows, as outset with the hypotheses. In order to grasp CSR as a (hegemonic) governmental tool in the GCCC, the dimensions of the variable are filled with elements from the consent and control framework. The components are deduced from the overall research interest and represent possible characteristics of the dimensions. Though, they only have a guiding function during the screening of the data for important information. Table 6.2 sets out the dimensions with their respective components.

Mediating actions: Activities linked to the implementation of CSR interventions
Mediating actions are all interventions and means of action which connect CSR as a transnational governance strategy with the resulting local transformations.

Mediating actions are defined for the given context as follows:

All strategic activities that a lead firm and its partners implement in order to achieve the goal of improved sustainability governance.

While these activities are directed at different levels of the chain, especially when it comes to cooperation and partnerships, in the following, only activities which target the local level of cocoa production are included the analysis.

Table 6.2 Independent variable "Transnational CSR intervention"—dimensions and components

Dimension	Components
Leading actors' (TNCs) characteristics	Interests Expectations/hypotheses Resource endowments Deficits Strategies
Other stakeholder characteristics	Stakeholder's role in CSR project Stakeholder's interests and expectations Sector responsibilities Interactions with other stakeholders
Scales of action	Administrative levels of activities Subjects of intervention per level Activities/mechanisms applied per level
Topic of CSR intervention	Agricultural practices Marketing system Sector policy design Social aspects Environmental aspects

Source: own elaboration.

What set of actions is applied in the course of the implementation of a given CSR intervention is subject of empirical study. Yet, from the knowledge obtained during desk study, some major fields of action are already known which are:

- The establishment of certification schemes,
- training of Good Agricultural Practicesn,
- material support for farm inputs and local infrastructure, and
- new forms of cooperation and partnerships.

The dimensions in Table 6.3 capture the most important aspects which are needed to understand the nature of the diverse mediating actions of TNCs when pursuing their consent and control strategy. A further partition into components is not useful here because the dimensions already capture the information that is needed to a sufficient extent. The dimensions are listed in Table 6.3.

Table 6.3 Dimensions of
mediating actions

Dimension
Actor
Subject matter of the action
Activities
Target group
Objectives (official, unofficial)
Character (formal, informal)
Resources

Source: own elaboration

Dependent variable "Ghana cocoa sector transformation"
As set out with the research hypotheses, the dependent variable "Ghana cocoa sector transformation" comprises two aspects of effects: the aspect of changing perceptions of sector stakeholders, particularly of cocoa farmers, linked to the introduction of CSR, and the changing roles and responsibilities of stakeholders at the local level of Ghana's cocoa production which lead to a new institutional configuration. The working definition of the dependent variable "Ghana cocoa sector transformation" is as follows:

In the present study, Ghana cocoa sector transformations linked to the intro-duction of transnational CSR interventions comprise changes in a) the perceptions of cocoa farmers' regarding their (future in) cocoa production and b) the local institutional environment of Ghana's cocoa sector.

Accordingly, the causal model's dependent variable "Cocoa sector transfor-mation" is subdivided into two parts, dimensions covering perceptions of cocoa farmers and dimensions directed towards transformations in the institutional environment. The perception part covers different aspects of farmers' percep-tions towards their cocoa farming such as their own agricultural capacities, their livelihood strategies, and major challenges and benefits of participation in the CSR intervention. The institutional transformations reflect three axes from the theoretical framework. Agricultural transformation and improved value chain organization are results of industrial governance efforts. The dimensions stakeholder alliances and distribution of tasks and responsibility are both impor-tant means of action and in the medium-term contribute to lasting institutional transformations.

As elaborated in Chapter 3, the establishment of the consensual status quo implies a sense of mutual interest in a way that subalterns agree to arrange themselves with their subordinated position and unequal resource distribution.

Shaping stakeholders' perceptions and attitudes in a way that they accept the dominant position of TNCs in the GCCC, and in our empirical case, especially in the Ghanaian cocoa sector, is regarded here as an important part of TNCs' governance strategy. Furthermore, from a TNC perspective, the attractiveness of cocoa farming needs to increase and modernization of cocoa farming has to appear as the common goal and mutual interest between poor cocoa farmers and transnational cocoa and chocolate industry. Looking through a neo-Gramscian lenses, in a situation of extreme poverty and the presence of other destabilizing factors which decrease the likelihood of maintaining the status quo, some concessions and improvements become necessary in order to prevent reorientation, if not uprising, of cocoa farmers.

Additionally, while pursuing the goal to efficiently intervene in cocoa policy-making, a status of a legitimate political partner becomes crucial and needs to be acquired. The importance of creating alliances to increase legitimacy and push the own interest has been described theoretically. From the consent and control-governance perspective appropriate organizational structures are recognized as fundamental governmental tools to achieve the above goals. Changes in institutional environments such as new types of cooperation between stakeholders, new mechanisms of introducing production norms, of developing capacities or of distributing important inputs, or the implementation of new types of control measures are all important components of institutional sector transformations linked to CSR. Table 6.4 shows the dimensions and components of the aspect a) the farmers' perception aspect and b) the institutional environment aspect of the dependent variable cocoa sector transformation.

Looking at the institutional side, as an important part of CSR strategies, certification schemes contain training on improved farming practices, labelled as "good agricultural practices". The dimension agricultural transformation of the dependent variable represents changes in agricultural practices and opportunities linked to the CSR intervention. Its components are the main aspects of cocoa production which are likely to be affected by certification schemes. For example, as has been discussed in detail in subchapter 4.4 and 4.5 on sustainability challenges in the GCCC, extensive farming practices or the use of child labor are problems in the chain that CSR interventions aim to address. Other problems cocoa farmers face with their production is the lack of input market access or the strict dependency on cocoa as their major cash crop. In the course of the project implementation, new mechanisms of resource flows, for both, inputs and outputs are likely to be implemented. Thereby, many new relationships might be established. The content of partnership agreements, the degrees of institutionalization and the distribution of key positions are important components which shed light

Table 6.4 Dependent variable "Ghana cocoa sector transformation"—dimensions and components

Dimension	Components
a) Farmers' perceptions regarding their cocoa production	
General perceptions regarding cocoa production	Interests, objectives/aspirations Strategies Attractiveness of cocoa farming Allegiance with LBC Satisfaction with living conditions Future plans
Agricultural capacities	Modernization Tradition Gender Self-estimation
Livelihood strategies	Cocoa farming as business (intensification) Extensification Diversification Migration Stagnation
Social and environmental practices	Tradition Gender Poverty Awareness
Challenges and benefits	Farming practices Input-output ratio Improvements
b) Transformations in institutional environment	
Agricultural transformation	Agricultural practices (incl. child labor, ecological practices) Inclusion into networks Access to infrastructure and inputs Market integration Subsistence farming
Alliances with sector stakeholders	Positions Topics of cooperation Tasks & responsibilities Degree of institutionalization Decision-making platforms (formal/informal) Channels of influence

(continued)

Table 6.4 (continued)

Dimension	Components
Value chain organisation	Control over production Control over resource flows Links between stakeholders Forms of information flow Forms of input distribution Marketing structures
Distribution of tasks and responsibilities	Functions created / positions established Forms of sector coordination and decision-making mechanisms Fora for policy dialogue

Source: own elaboration.

on the institutional effects of CSR. Similarly, organizational patterns of the sector are affected. New actors enter the picture with new responsibilities, endowed with different resources, pursuing their own agendas and interests. Their respective activities in turn shape activities of already existing actors in the field and new forms of interaction and cooperation might occur. In this dynamic situation, norms and values regarding cocoa production and the institutional environment are likely to spread and new forms of sector coordination evolve. The empirical study aims to capture these dynamics.

6.3 Field Work Strategy: Triangulation of Qualitative Research Tools

The empirical study consists of two main parts: an explorative study and a case study on one case of cocoa sustainability certification in Ghana. The explorative study served to grasp the field, that is to get an improved understanding on the sector composition and existing CSR interventions, the different forms of implementation at the local level, and to identify the main stakeholders. Based on the findings on sector constellations and the diverse CSR approaches, the next step was deciding on one CSR project that would be appropriate for the detailed study of local institutional transformations and farmers views linked to transnational CSR. A field work was designed and the respondents were mainly cocoa farmers which were targeted by the CSR project selected for case study. These CSR-targeted cocoa farmers were interviewed about the perceived effects of these interventions on their cocoa production and livelihoods, but, in order to contrast the picture (but not to compare), non-CSR-targeted farmers and other relevant sector stakeholders were included to the field work, too.

In both stages of inquiry, the triangulation of qualitative research instruments sought to obtain a broader and more context sensitive understanding of the field. While semi-structured key informant interviews with different degrees of openness have been the main source of data collection, focus group discussions and non-participant observations as well as interaction documentation were also important sources of information during the course of the research process. In the following, the design of both research sequences is described in detail.

6.3.1   Study of Sector Constellation and Existing CSR Interventions in Ghana's Cocoa Industry

In order to obtain a deeper understanding of the functioning and ongoing transformation processes within Ghana's cocoa sector, semi-structured key informant interviews were conducted with sector representatives in Ghana (62 key informants) and Côte d'Ivoire (7 key informants) in 2015 and 2017. The interviews in Côte d'Ivoire were aimed at gaining a better understanding of sector dynamics by contrasting the Ghanaian field and thereby grasping its particularities but were not included the data analyzing process. Furthermore, out of the 62 key informant interviews in Ghana, 26 were included the data processing whereas the others served as background information. The sampling was purposely executed in a step-wise way based on the information obtained in the previously conducted interviews, known as the snowball system.

The purposeful stepwise sampling was derived from Grounded Theory methodology developed by Glaser and Strauss (1967). It accounts for the openness in the field to include (ideally) all necessary cases into the sampling which are identified during the research process and to readjust the research tools if needed. The aim is to obtain the widest possible range of opinions and experiences with CSR in the Ghanaian cocoa sector. The explorative study comprised members and representatives from the public, the private, and the NGO sectors. From the public sector, staff members from COCOBOD and its several subsidiaries, especially CHED, local government agents, and representatives from development agencies and donors were interviewed. From the private sector, most notably, interviews were conducted with sustainability officers from the LBCs, PCs as well as staff of transnational processing and manufacturing companies, and the WCF. In addition, representatives from various institutions such as from

NGOs, certification organizations, Ghanaian universities, and farmers' associations were interviewed. This wide breadth of expert opinions helped to gain a solid picture of most of the contrasting perspectives in the field which in turn enabled the researcher to identify differences and commonalities regarding the experiences with the implementation of CSR in the sector. The semi-structured interviews contained questions regarding the following topics:

- General overview of the Ghanaian (and Ivorian) cocoa sector functioning
- The role of stakeholders, their interests and instruments
- A pro-active governance role of TNCs by implementing CSR
- Decision-making processes and bargaining positions
- Restructuring/transformation of the sector governance
- Personal evaluation/legitimatory effects

The interview guide as well as the list of interviewees are documented in the Electronic Supplementary Material. The research authorization was provided by COCOBOD, the public institution in charge of all cocoa regulatory affairs in Ghana.

6.3.2   Farmers Interviews on Experiences with the Selected Sustainability Intervention

The survey among CSR-targeted farmers aimed to capture the effects of one selected CSR initiative in the Ghanaian cocoa industry. As set out in the causal model, the study considers several levels of CSR effects: farmers' regarding their cocoa production, perceived effects of the intervention on the livelihood conditions of the participating farmers, perceived changes in agricultural practices and sector functioning. In order to capture these aspects, a set of multiple research tools has been applied. In the present context, data was collected through pre-structured, guideline-based key informant interviews (Gläser and Laudel 2012, p. 111). The relatively open design of the interviews, which had a conversational character, made it possible to learn about the individual assessments of the interviewees. Such a semi-standardized interview is particularly suitable for addressing not only the explicit assumptions of the interviewees, on the basis of which open questions can be answered spontaneously, but also their implicit assumptions (Flick 1995, p. 99ff.). For this purpose, supplementary questions were asked based on the author's previous knowledge from desk research and research interest. Complementing the open questions with these hypotheses-based ones made it possible to obtain concrete information on the implementation process of CSR in Ghana in addition to personal assessments and opinions.

The interviews consisted of several thematic sections, starting with open questions about the overall assessment of the interviewees' situation as a cocoa farmer, their personal story of their cocoa farming, and the associated problems of being a cocoa farmer. If the interviewees did not mention specific aspects which were relevant for the causal model, more precise, hypotheses-based questions were asked. The interviews were supplemented by other research instruments such as focus group discussions with farmers' societies, non-participatory observations in farmer trainings and group meetings, as well as interaction documentation and were applied depending on the context. During the two study sequences in 2015 and 2017, a total of 56 targeted and non-targeted cocoa farmers by one case of CSR intervention were interviewed and five focus group discussions conducted. The objective was to identify *tendencies* of changing perceptions over the two years of participation in a CSR project.

A three-step sampling process
The sampling process comprised of three steps. In a first step, one CSR intervention had to be selected for the study. In a second step, within the area of this particular CSR intervention, project zones had to be chosen for the sampling of interviewees, the final step of the sampling.

a) Selection of the case
Among the multitude of transnational CSR initiatives in Ghana's cocoa sector, one important CSR initiative has been chosen. As highlighted above, in order to guarantee anonymity of interviewees, the name of the studied project and participating companies cannot be made public. The studied project was purposively selected for the case study for several reasons. The cooperation between one of the lead processing companies and an important local LBC represented a new pattern of interactions within the value chain which began evolving with the new type of CSR projects in the first half of the 2010 s. At the same time, it was one of the most complex programs. As a result, diverse new roles and responsibilities evolved from the program. It is therefore an interesting case to analyze and discuss the institutional changes in the sector. The contributing partners were among the most active private actors in the Ghanaian cocoa sector and played a major role within the transformation process of the sector. Additionally, the initiative

was implemented in the Western Region[1], the region with both, the highest share of cocoa productivity and with the highest rate of food insecurity among cocoa farmers. The next subchapter introduces several key pieces of data on the project area. Surprisingly, when returning to the field in 2017, the studied project has been abandoned and did not exist anymore but this information will be treated in the case-study part and is part of the study results.

The prior decision of the case led to a mixed sampling procedure for the interview partners which is described hereafter.

b) Area of the study
The selection process of the communities and survey participants combines purposeful and random sampling elements. The CSR project was located in the Western Region of the country. The project administration was divided into the three zones A, B, and C with each having a certain number of operational districts. As a first step, one zone was selected randomly. The selected zone consists of six operational districts, each having up to 50 target communities. The purpose was to include different types of communities with different characteristics in the sample. Therefore, the communities from the six operational districts were classified as follows:

(1) very remote and difficult to access,
(2) remote but good access to road, or
(3) near town, close to main road.

In a next step, a random selection of five communities out of the classified pool was done: two communities were chosen from type (1), another two from type (2), and one community from type (3).

c) Sampling of interview partners
The sampling of the participants in the survey was purposely aimed at reflecting the composition of the farmer society members in each survey community. A sensitive community entry strategy was designed. During the first visit, a meeting with members from the studied project was arranged with the help of the PC from the partnering LBC and the COCOBOD community extension agent.

Besides getting to know the farmers group and establishing contact to possible interview participants, these initial meetings fulfilled the following objectives:

---

[1] The region has been divided and since then, the survey area is under Western North Region, see Section 6.4.

- First, they aimed at informing the farmers about the study, its research objectives, its independency from any other institutions and the conditions of participation.
- Second, a focus group discussion was conducted with participants in the project and some non-participating farmers from the same communities who were attending the meeting. All members of the studied project LBC society of the respective community were invited to participate in this meeting. During these meetings, the author gained an understanding of what matters for the farmers what helped her to finetune the questions for the individual interviews.
- Third, the objective was to identify farmers at the end of the meeting who were willing to participate in the study and to register for the interviewee pool.
- Fourth, the aim was to collect some socio-demographic information from those farmers who registered for the interviews. This information was needed to conduct the purposive selection procedure. In addition, they were asked to leave their contact data in order to come back to them after the selection procedure had been completed and inform them on the results.

Based on this information, the purposeful sampling of the interview partners sought to achieve a group of interviewees which covers all important positions and was done according to the following criteria:

- Select crucial positions: chief farmer, lead farmer, Purchasing Clerk
- Select 5 representative farmers from the project LBC (2 females, 3 males = > selected randomly from a wide range of age and contrasting strength of resource endowments, fewer females because of their lower presence in the studied project societies)
- Select 2 sharecroppers from the project societies (all participating sharecroppers were male)

During the research process and in accordance with the methodological approach of aligning sampling with the findings in the field, the need to include non-members of the project society in order to contrast position became more evident. Therefore, a number of them have been included in the sampling at community level. After the selection, the participants were called and meetings were arranged. The Table 6.5 summarizes the sampling procedure applied for the study of farmers' perceptions.

Table 6.5   Farmers' perceptions study: sampling procedure

Selection item	Selection result	Procedure	Explanation
Case	the studied project	Purposively	Representative and interesting example of CSR in Ghana's cocoa sector
Study area	Western Region, Sefwi District	Purposively	Area of the studied project intervention
Population—communities	Project-targeted communities	Randomly	5 communities out of 35 target communities
Population—interview respondents	project-targeted farmers and share-croppers, members of project society in Sefwi Wiawso district, non-project-targeted farmers	Purposively	67 cocoa farmers interviewed in the study period from 2015–17, selected in order to represent the project society-population in the respective community

Source: own elaboration.

Methods and content of the survey

The main tool for information gathering was the semi-structured interview with the selected interview partners. The interview guide was designed in order to reflect the main research interests of the study. The questions were drawn combining elements from the theoretical guiding framework outlined above as well as from the insights gained during the explorative phase and the focus group discussions.

The following topics provided the frame for the interview guide:

- Personal story of the interview partner as a cocoa farmer (including experiences with major changes over the past decades in the area of cocoa farming)
- Experiences with the participation in the studied certification project (including reasons/incentives for participation and how these are met, and the difficulties and benefits)

- Perceived effects of the certification project on cocoa production and future as a cocoa farmer
- Livelihood changes linked to the introduction of the program

The detailed interview guide is provided in the Electronic Supplementary Material. The mean length of an interview is of approximately one hour. While only a few interviews took less than half an hour, many of them took a bit longer, up to 90 minutes in some cases. The interviews took place either at the houses or farms of the participants or in the case these were too far at the cocoa depots of the project LBC, often close to the purchasing clerk's house. The author acknowledges the loss of ethnographic insights and information coming with the decision to conduct the interviews at the cocoa depots. In addition, having the interview at the depot bears the risk that interviewees would adjust their responses more to what they think would be good for the project LBC rather than the response they might give when asked at their own home. Financial and time resource restraints did not allow to visit all farmers individually.

Other methods of data gathering included the focus group discussion (FGD) and the means of non-participatory observations. The latter was realized in cases where the author gained the opportunity to participate in a farmers' certification training or a lead farmer training. The notes which were taken were included in the research process and strongly enriched the authors' understanding of informal power relations, expectations, and (mis-)understandings between stakeholders.

The FGD were arranged as described under the section c) selecting interview partners. The number of participants differed highly between the communities: in most communities the number of participants was about 20 but in one community about 60 participants came together for the meeting. While the discussions with 20 participants were still relatively easy to moderate, the discussion with the 60 participants was a challenge, but nevertheless brought a number of interesting insights. The difference in the numbers are linked to the composition of the communities and the way the certification project is organized in them. The smaller the community, the smaller the project societies. In the case of very small communities, communities might be combined and then comprise of a high number of members. In all cases, the community extension agent or the certification agent introduced the author and the study to the group while then the interpreter took over.

6.4 Data Processing and Analysis

The analysis process is in line with the approach of the qualitative content analysis method by Gläser and Laudel (2012). Initially, content analysis was developed as a tool to quantify information from texts and to allocate it to a rigid theoretical grid. In the German social science context, a qualitative approach of content analysis was established and goes back to Mayring who further developed the content analysis in the 1980 s. While still sharing the main element of quantitative content analysis to extract information from a text and treat it separately to it, he introduced aspects of qualitative research to it. By doing so, Mayring still uses a rigid set of theoretical categories which manifestations are determined based on theoretical considerations and to which the empirical information has to fit (Mayring 2015). Gläser and Laudel modified Mayring's approach to qualitative content analysis by opening it to changes and widening during the process of analysis. The main activities of this form of content analysis are extraction of empirical information, the processing of the data, and the evaluation and interpretation of the data. All steps are based on the researcher's individual decisions but a set of extraction rules increases the congruency between decisions and the replicability of the single steps. Qualitative content analysis differs from the Grounded Theory procedure, which is an often-applied approach for the analysis of interviews and other empirical data documents and which uses techniques of coding of information and synthesizing it into categories and finally a ground-based theory. In contrast to this, the process of extraction is an activity where the information which is relevant for the research context is extracted from the text and allocated to a theory-based set of categories but which is open for modification if the empirical data demands it. The evaluation grid is made from variables' and mediating action's definitions, dimensions, and components. Thus, this approach supports the researcher to remain closely in line with the overall research interest but leaves room for modifications based on empirical evidence.

This approach to qualitative content analysis led to the development of the evaluation grid for data extraction from the interviews which is based on the above-presented causal model of CSR-induced cocoa sector transformation in Ghana. During the whole data assessment process, the model guided the analysis and information that was extracted and allocated with the already existing categories which are founded in the dimensions and components of the variables and mediating actions. In the Electronic Supplementary Material, an overview on the final set of categories can be found.

During the data processing, the various forms of data have been treated differently. While all interviews were included the analysis and evaluation phases, the memos from the observations and field interactions were used only during the interpretation phase. The analysis was conducted with the help of the commercial software MaxQDA. The lists 1 and 2 in the Electronic Supplementary Material give an overview over all conducted interviews.

6.5   Characteristics of the Research Area and Socio-demographics of Interview Partners

The selected project, similar to many other CSR projects in the Ghanaian cocoa sector, was first implemented in the most productive cocoa area of the country, the Western region. According to the findings of COCOBOD's Project Coordination Unit, the Western Region is the region most targeted by transnational CSR interventions. The most likely reason that this area has a high number of interventions is due to its high density in cocoa production (see Section 5.1). Ghana's last cocoa frontier in the Western tropical rainforests has expanded since its opening in the 1970 s and over the past two decades farms with intensified cocoa cultivation spread.

In February 2019, following a referendum in 2018 on the creation of six new regions, the Western Region was divided and the northern part of the region became an individual region, the Western North Region.

All communities which were included in the survey are located in the area of the Western North region. The young region consists of nine districts: Sefwi-Wiawso, which became the capital of the region, Aowin, Bia East, Bia West, Sefwi-Akontombra, Juaboso, Bodi, Suaman, Sefwi-Bibiani/Anhwiaso/Bekwai. After Sefwi-Wiawso, Aowin and Sefwi-Bibiani/Anhwiaso/Bekwai have the highest population. To the west, the region shares its border with the Ivory Coast, and to the southeast and north with the Ghanaian regions Central Region, Ashanti, and Bono.

Due to the recent nature of the split, at the point of time of writing, there is almost no data available which only concerns the newly created region, but there is little information on the demographic and economic situation available. Data that could be found at the time of writing still concerns the situation in the former Western Region prior to the division. This data, however, likely does not properly represent the situation in Western North. For instance, in the Ghana Living Standards Survey Round 7 on poverty trends in Ghana between 2005 and 2017 (Ghana Statistical Service 2018, p. x), the Western Region is described as one of the five regions in Ghana which shows less prevalence of poverty than the rest of the country. But this might not hold for the northern part of the region.

There is a strong imbalance between the northern and southern areas within the former Western Region: the prevalence of formal and (mainly) informal business establishments is unequally distributed within the region with a developmental advance of the southern part of it. While over one-quarter of business establishments are located in the large city of Sekondi-Takoradi, which hosts the second most important harbor in the country, the southern and central parts of the region are more economically diversified than the northern part. Frontier dynamics are indicated by the higher share of rural population in Western North compared to the rest of the Western region. According to the Population and Housing Census from 2010 (Ghana Statistical Service 2012, p. 21), 58% of the population in the Western Region live in the rural areas and 42% in the urban settings (compared to 49% and 51% respectively in the whole country). However, taking a closer look at the population numbers from all districts which became the Western North region, the picture looks different here, too: 563.562 people live in the rural area and 146.873 in urban settings, making it 80% of the population in the Western North who live in the rural areas.

As already was the topic in Chapter 5, many people from the early cocoa regions in the Eastern and middle sides of Ghana as well as from Northern regions and countries migrated to the Western region to look for their future in and around cocoa farming. In the 2010 census (Ghana Statistical Service 2012, p. 34), the largest group in the Western region are Akan people, including their different subgroups such as the Sefwi who are autochthon inhabitants in the region. The other two largest groups in the region are Mole-Dagbini from the north and Ewe from the east, giving evidence of the two major cocoa migration streams (cocoa farmers from the east looking for new land when their soils became less fertile, and mainly unemployed poor people from the north looking for entering cocoa farming activities). The main economic activities in the rural area of the region are agriculture, mining, and logging. Agricultural production includes cash crops, mainly cocoa and rubber, and staple crops produced for the local markets as well as for own consumption. The mining sector comprises of both, large legal mineral mines (e.g. gold and bauxite but also other rare earths), and the rapidly spreading illegal superficial gold mining which drastically destroys the local landscape and is seen as the main competitor with cocoa production (see Snapir et al. 2017 on Galamsey in Ghana). Logging also exists in both legal and illegal forms and still poses a threat to the remaining rain forest reserves (Ametepeh 2017) in the region. In the urban centers of the region, many different formal and informal services provide income opportunities to the dwellers.

The communities, in which the research activities were conducted are located in the following four districts in the Western North region:

- District Juabeso: one very remote community (Juabeso_Type1)[2], one not so far from main road (Juabeso_Type2)
- District Bodi: one community not far from main road (Bodi_Type2)
- District Sefwi Wiawso: one community located at the main road, close to main town (Wiawso_Type3)
- District Akontombra: one very remote community consisting of three settlements (Akontombra_Type1)

The map 6.1 shows the areas of the research and map 6.2 the approximate locations of the communities.

Map 6.1 Research area in Western North region, Ghana. (Source: map prepared by the author. Map data: mapz.com 2020, OpenStreetMap (ODbL))

Interviews were conducted in five communities in the Western North region according to the sampling features described above. Juabeso_Type1 is a small community which has about 150 inhabitants and is located approximately 20 km

---

[2] The communities and the farmers' groups to which the interviewees belong to are very small and it would be possible to trace statements back to individuals. Hence, in order to assure anonymity of the interview partners, the names of the communities are replaced by the codes; the types refer to the sampling categories introduced in Section 6.3.2.

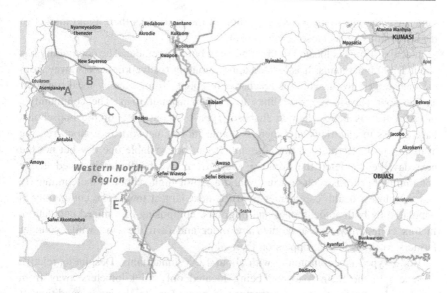

Map 6.2 Approximate locations of the selected communities. ((A = Juabeso_Type2, B = Juabeso_Type1, C, Bodi_Type2, D = Wiwso_Type3, E = Alontombra_Type1) Source: map prepared by the author. Map data: mapz.com 2020, OpenStreetMap (ODbL))

away from the next bigger town.[3] Due to the poor state of the clay feeder road (see pictures in the Electronic Supplementary Material for examples of poor roads in the region), it takes about 50 minutes to get to the town by local transport means. The community is surrounded by rainforest, and all of its inhabitants are migrants or descendants from migrants which settled there because of cocoa production. More than the half of the community members are young farmers who are part of an extended family network and therefore hold shares of the farm lands. In 2017, there was no electrical grids and no mobile phone network available in the community and drinking water had to be fetched from the stream. A primary and a junior high school are located in the closest smaller town, which is about five kilometers away. The next health center is only available in the next bigger town, so people have to travel about one hour for medical attention. The

---

[3] All information about the community characteristics are obtained from expert interviews with key informants or the farmers living there. It is important to note that all figures are estimates that interviewees have made. Even if some numbers appear as precise figures, they cannot be taken as official demographic data.

same distance has to be travelled to shop at the market as there is no market in the community or its closer environment.

Juabeso_Type2 is located in the same area but less remote. The community has about 1,200 inhabitants. It is located approximately twelve kilometers away from the next largest town, which is reachable in about 20 minutes by local transport as the poor clay feeder road links to a main asphalt road. The community's infrastructure is better than in Juabeso_Type1. About 70% of the households have access to electricity and there is stable mobile network reception. Yet, the availability of electricity is not constant. It can be available one day but not the next day. People fetch the drinking water from two boreholes installed in the community. There is both, a primary and a junior high school in the community and the closest health care center is in the next largest town. The community is populated by both autochthones and migrants, composing of a mixed population. In this community, cocoa farmers are older and own the land while younger farmers work as share croppers on their plots.

Bodi_Type2 is a community with about 1,000 inhabitants. Despite its relative closeness to the main road and being located only 15 kilometers away from the next largest town, its infrastructure is poor. Due to the poor condition of the clay feeder road, it takes about 30 minutes by local transport to reach the next largest town. There is no electricity and no mobile network available, and people have to fetch their drinking water from a stream. Dwellers use the same electricity substitutes as in the community Juabeso_Type1. There is no school in the community itself but in the next smaller town which is two kilometers away. The next health care center is about one kilometer away. There is no market in the community but in the smaller town where the schools are located. The people living in Body_Tpye2 are mainly autochthonous to the area. Most of the cocoa farmers are older farmers who own the farm land, most of the younger farmers work in share cropping arrangements on the older farmers' farms.

Wiawso_Type3 is the community with the best infrastructure out of the five communities. Being located directly on an asphalt road and only seven kilometers from the next largest town, reachable in ten minutes by local transport means, the general situation of the community is distinctly better than in the remote areas. In Wiawso_Type3, about 95% of the inhabitants have access to electricity and the mobile network is stable. People fetch their drinking water form one borehole. There are a primary and a junior high school located in the community and the next health center is one-kilometer away. The next market is located in the next bigger town. The community has a population of about 950 with mainly autochthonous inhabitants. In this community, through inheritance, about half of the old farms have been transferred to the younger inhabitants.

Akontombra_Type1 is the most difficult to access out of the five communities due to the very poor clay feeder road. The 300 inhabitants-comprising community is located ten kilometers away from the next largest town, but due to the extremely bad road condition, it takes about 50 minutes to reach it by local transport means. Approximately 30% of the inhabitants have access to electricity and the mobile network reception is poor. There are two boreholes in the community for water. The distance to the next health center is nine kilometers which implies travel time of almost one hour. There are primary and junior high schools in the community but the closest market is only located in the next largest town. The population living in Akontombra_Type1 is mixed, comprising of autochthonous inhabitants and migrants. The difficulties to access the community are reflected by a larger share of young farmers having their own cocoa farms since more land was still available that could be purchased by the migrants. This, however, is a typical pattern in Ghanaian cocoa farming and a feature of frontier movements as described above already. Farmers keep on looking for possibilities to move to more remote areas in order to find richer soils to produce their cocoa what becomes increasingly limited. Therefore, cocoa farms in the remote areas are generally younger than in more accessible places. In four out of five communities, the project's farmers' society to which most of the interview partners belong to, comprise of about 40 members. Only in Akontombra_Type1 the group is larger with more than 60 members. The societies are composed by more than the half of male and less than the half of female participants but in the initial meetings, women were still clearly underrepresented.

As can be seen in Table 6.6, many more men were interviewed than women (40 and 16 respectively). This is due to the low representation of women in the introductory focus group meetings, especially in these communities where only a few women are included in the sample (Juabeso_Type1 and 2 as well as Bodi_Type2) and the fact that they almost do not hold any key positions such as lead or chief farmers or PCs. Most of the interviewees are members of the project LBC's society and participants of the certification project at the same time but this does not hold for all respondents.

The Table 6.6 on the characteristics of the interview partners illustrates that despite their difference in accessibility and remoteness, the communities do not differ drastically in their characteristics. The age of all respondents ranges from 25 to 78 but age information was not collected in all communities.[4] Therefore, the indicator "years engaged in farming: mean" shall provide better information about the composition of the interviewees per community. It is highest in the Bodi_

---

[4] For the detailed interviewees data see appendix 9.

Type2 community and lowest in Juabeso_Type1 and Akontombra_Type1. This, however, is an indication that the more remote the communities have younger farms, not necessarily farmers, and that these areas have been explored later when no more fertile lands in the less remote areas were available. The number of household members for all five communities ranges between six and eight people and seems to represent a typical household in the area while the average household size in the country is 4.4 with a declining trend (Ghana Statistical Service 2012, p. 3). The total mean number of plots is three but in four communities it is less than three. Only in the community Body_Type2, farmers have more than four plots in average. Bodi_Type2 stands out in two other factors: one, it also has the highest number of years engaged in cocoa farming, and two, together with Wiawso_Type3, respondents supplement their cocoa farming with other income sources less than in the other areas. The indicator "other income sources" is highest in the two remote communities Juabeso_Type1 and Akontombra_Type1. The data of the individual respondents of these two communities[5] shows that "farming other crops" is the most important additional income source for cocoa farmers there. Looking at the responses to the question of engagement in "other crop production" in the five communities, it is highest in Body_Type2 and Akontombra_Type1 followed by Juabeso_Type1. Figure 6.2 gives an overview on the main community characteristics.

---

[5] See appendix 9.

Table 6.6 Summary of interview characteristics per community

Community code	Number of interviewees	Positions: sum 1 = AA Lead farmer 2 = AA PC 3 = AA cocoa farmer 4 = cocoa farmer others 5 = Chief (AA-farmer) 6 = Chief (non AA-farmer) 7 = AA sharecropper	Sex: sum f/ sum m	Age: mean	Household members: mean	Household head yes: sum	Member of AA group yes/no: sum	Participant certification project yes/no: sum	Number of plots: mean	Years of cocoa farming: mean	Number of other crops: mean	Having laborers yes: sum	Other income sources yes: sum
Juabeso_Type2	8	1 = 1 2 = 1 3 = 6	f = 2 m = 6	n/s	8	8	yes = 8 no = 0	yes = 8 no = 0	2,9	20,1	2,3	6	5
Juabeso_Type1	10	1 = 1 2 = 1 3 = 5 5 = 1 7 = 3	f = 1 m = 9	50	6,7	7	yes = 10 no = 0	yes = 10 no = 0	2,7	17,7	3,7	5	7

(continued)

Table 6.6 (continued)

Community code	Number of interviewees	Positions: sum 1 = AA Lead farmer 2 = AA PC cocoa farmer 3 = AA cocoa farmer others 4 = cocoa farmer 5 = Chief (AA-farmer) 6 = Chief (non AA-farmer) 7 = AA sharecropper	Sex: sum f/ sum m	Age: mean	Household members: mean	Household head yes: sum	Member of AA group yes/no: sum	Participant certification project yes/no: sum	Number of plots: mean	Years of cocoa farming: mean	Number of other crops: mean	Having laborers yes: sum	Other income sources yes: sum
Bodi_Type2	8	1 = 1 2 = 1 3 = 3 6 = 1 7 = 2	f = 1 m = 7	n/s	6,3	9	yes = 7 no = 1	yes = 7 no = 1	4,3	22,2	3,9	4	4
Wiawso_Type3	11	1 = 1 3 = 5 4 = 4 7 = 1	f = 4 m = 7	46,2	6,1	5	yes = 7 no = 4	yes = 4 no = 7	2,7	20,5	2,8	2	4

(continued)

Table 6.6 (continued)

Community code	Number of interviewees	Positions: sum 1 = AA Lead farmer 2 = AA PC 3 = AA cocoa farmer 4 = cocoa farmer others 5 = Chief (AA-farmer) 6 = Chief (non AA-farmer) 7 = AA sharecropper	Sex: sum f/ sum m	Age: mean	Household members: mean	Household head yes: sum	Member of AA group yes/no: sum	Participant certification project yes/no: sum	Number of plots: mean	Years of cocoa farming: mean	Number of other crops: mean	Having laborers yes: sum	Other income sources yes: sum
Akontombra_Type1	19	2 = 1 3 = 17 4 = 3	f = 8 m = 11	42,4	6,8	12	yes = 17 no = 2	yes = 17 no = 2	2,5	16,8	3,9	6	7
Total	56	1 = 4 2 = 4 3 = 36 4 = 7 5 = 1 6 = 1 7 = 6	f = 16 m = 40	46,2	6,78	44	yes = 49 no = 7	yes = 46 no = 10	3,02	19,5	3,32	23	27

Source: own elaboration.

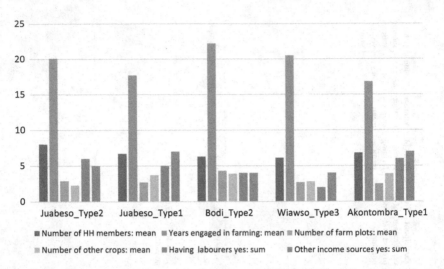

Figure 6.2 Characteristics of interview partners per community. (Source: own elaboration)

CSR in Practice: Assessing the Implementation of one Cocoa Sustainability Program in Ghana

7

In this chapter, the results of the empirical study on transnational CSR in Ghana's cocoa sector are presented. The case study project, hereafter "the studied project" has been selected in order to shed light on the implementation process of transnational CSR and resulting transformative dynamics in Ghana's cocoa sector. The studied project was a tripartite partnership project which sought to improve sustainability of cocoa production in Ghana by implementing the UTZ Code of Conduct and certifying achievements by UTZ certified. The two main aspects of the project were the improvement of extension services and the establishment of a cocoa traceability scheme. The studied project represents an instructive case of how TNCs liaise closely with other sector stakeholders in a given specific local sector constellation in order to establish their CSR and sustainability strategies. For the TNC running the studied project, it was the main project in Ghana under which the company implemented its overall cocoa sustainability strategy. The project partners were one of the LBCs in Ghana and an international NGO. Having started in 2012 and being a successful tool to spread GAPs and other UTZ standard requirements in Ghana's last cocoa frontier, the joint venture ended abruptly in 2016 due to financial struggle of the partnering LBC. The split of the field work into two phases provided the author with the valuable opportunity to study the transnational processor's reaction to the shock and to follow how the company seeks to further its CSR strategy under the new circumstances. The resulting dynamics are unprecedented in Ghana's cocoa sector and reveal how intensely lead firms from the GCCC seek to amplify their influence on the production level under the umbrella of their sustainability interventions.

The chapter begins by providing a detailed insight in the stakeholders' motivations for and positions in the project and its process of implementation. This part

F. Ollendorf, *The Transformative Potential of Corporate Social Responsibility in the Global Cocoa-Chocolate Chain*, (Re-)konstruktionen – Internationale und Globale Studien, https://doi.org/10.1007/978-3-658-43668-1_7

183

of the present study exposes the mechanisms behind the transnational CSR intervention in the form of an LBC model of sustainability certification in Ghana's cocoa sector and corresponds to the guiding model's element "mediating actions". The Sections 7.2 and 7.3 present the findings regarding the two aspects of the transformative process which are of major interest for the present study: changes in cocoa farmers' perceptions regarding their cocoa production, and transformations in the institutional environment of Ghana's cocoa sector. Both changes are analyzed in the light of the multitude of similar CSR interventions implemented in Ghana's cocoa sector during the same period. The findings provide the basis for the discussion of the research questions in Chapter 8.

7.1 The Implementation: Modus Operandi of the Studied Project

7.1.1 Project Partners

Project partners were the studied transnational cocoa processing and global food corporation (hereafter "the project lead firm"), a major Ghanaian LBC (hereafter "the partnering LBC") and an international NGO (hereafter "the partner NGO"). Other cooperating partners to the project were UTZ certified and COCOBOD. The main drive of this stakeholder constellation was the project lead firms' interest in certified beans but inability, due to its inexperience with local structures, it was not able to implement the needed structures on its own. It was the partner NGO then who established the link between this strong LBC and the transnational processor to implement a project of UTZ certification. UTZ as the standard provider and certifier assisted the project team with trainings of head level officers who then triggered the knowledge down to the rest of the staff. COCOBOD became a partner through the CEPPP and provided its qualified extension agents to the project.

The project lead firm
The project lead company is among the world's most powerful global food corporations and It engages in agricultural products and services in the fields of animal nutrition, food and beverages, bioindustry, food services, risk management, meat and poultry, beauty, pharmaceuticals and transportation. Being one of the leading producers of food ingredients such as starch, glycose syrup or vegetable oils, the processors' primary commodities are inside a multitude of brand products but not visible to consumers (Ayoub n.d.a). Among its biggest customers

are Nestlé, McDonalds, Burger King, Walmart, Kellogg's, Unilever and Danone (Ayoub n.d.b). Hence, cocoa processing is only one of the many branches of the company but that does not imply that it would not be a leading company in the sector. The company does have cocoa processing plants in all strategic locations: in Africa in Cameroon, Côte d'Ivoire and Ghana, in Asia in China, Indonesia and Malaysia, in Europe in Belgium, France, Germany, the Netherlands and the UK, in America in Brazil, Canada and the USA.

The company is engaged in a big number of partnerships and sustainability multi-stakeholder initiatives. But what can be observed in the GCCC seems to be different from other sectors. The interest in CSR and sustainability interventions here is more inherent to the company's business strategy. As has been elaborated in the introductory chapter, the dependency on the willingness of so many small-scale producers to continue cocoa production fosters an intrinsic motivation to achieve a real improvement of local conditions. Hence, the drive here is different to the one in soy or palm tree production which is based on large-scale producers. The following statement by one representative of the project lead firm elucidates this interest.

> "Because we believe that if the farmer is well off in terms of his farming activities but the community he lives in is short of infrastructure then the farmer won't be happy to live in that community to farm and might be willing to leave, to come to Accra or Kumasi or any big city because of various good reasons. There is no good drinking water, no healthcare and things like that so the farmer would not be happy." (Project lead firm representative in April 2015)

And next to this concern which explains the TNC's simple need to do something for the improvement of farmers' livelihoods, its major business interest to increase quality and quantity of cocoa bean production, hence the particular interest that needs to be mainstreamed (see Subchapters 3.2 and 3.3), clearly is the other main driver of its CSR and sustainability interventions:

> "Yeah, our motivation is to see increasing yield. But potentially, Ghana is producing 350kg to 400kg per hectare compared to other nations that are doing 800kg to 1500kg. So we realized that there is potential in Ghana and a need to increase the yield. So the motivation is to see an increase in yield, this is why we bring the training, access to inputs and all that to the farmers so that we see them intensifying production and not just extensifying and going to forest areas but on the same piece of land to increase and double the production. So that is the first motivation about it. And then also to ensure that there is sustainability in cocoa. So we have cocoa today and tomorrow." (Project lead firm representative in April 2015)

With this statement, the interviewee makes no secret of the fact that the studied project serves the business objectives of the company to modernize the sector. It seems normal to him to talk about the farmers as if they were only an important instrument of these business objectives and that they must therefore function as the company needs them to.

These two statements of the project lead firm's staff member give insight in the main interests of the TNC: one, the increase of productivity, and two, to ensure supply sustainability. Therefrom derived objectives are to make farmers remain in the sector, and to establish socially and ecologically sound practices which allow production sustainability, as well as the modernization of farming practices.

The project lead firm has initiated its cocoa CSR program in 2012 and since continuously further elaborated to a highly sophisticated strategy. Conducting research on many aspects of cocoa production, the company has identified a multiplicity of issuesKlicken oder tippen Sie hier, um Text einzugeben. which are regarded to be of key relevance and which also showed the highest correlation between stakeholder interests and the company's business success. These topics were then chosen as high priority areas of cocoa CSR strategy and are the following: living income, human rights and child labor, community wellbeing, deforestation, supply chain transparency, and farmer profitability.

During the time of the first field work sequence, 2015, there were still only three main thematic pillars of the strategy: farm development, farmers training and community support. The studied project was implemented under the pillars farmer training and farm development.

Different to the other countries, in Ghana, the project lead firm was confronted with the system of internal marketing and COCOBOD's strong position in the sector. In contrast to the other countries, it did not have any direct marketing links with cocoa farmers and was therefore lacking the important basis of target farmers. Therefore, the company needed to establish a cooperation with other sector stakeholders in order to implement its cocoa CSR program in Ghana. As highlighted in the following statement, the studied project was therefore a perfect project for the company to overcome this situation.

"So in Ghana the market is liberalized. So the LBCs they have their direct relationship with the farmers. So it is good to have a good relationship with some of the LBCs so you can always have access to the farmers and then get the data as soon as you want it." (Project lead firm representative in April 2015)

Nevertheless, the direct implementation activities on the ground were rather assigned to the partnering LBC's sustainability staff while the project lead firm was the main funder of the activities. Its sustainability staff sometimes went to the field, too, but was rather involved in training of key resource people such as lead farmers and project staff. Together with the partnering LBC's sustainability manager, the the project lead firm's sustainability officer oversaw the progress of the project and reported back to the headquarters which sought to gather information from the local levels for further strategic development. The interviewees' statements on future visions regarding Ghana's cocoa sector seem to give some insight on the general vision when it comes to the professionalization of farmers—or the overall sector: The way forward in the sector would lie in the creation of specialized work forces who conduct important sector services and the pooling of small-scale farms to big farms which are managed by professionals. But this vision seems to be still some time ahead. For now, that is the 2020 s, as stated in its cocpa CSR program description, the strategy is to boost small-farmers productivity while integrating them into a net of services which is needed for a sound production and to improve their living environments to keep them in production.

The international partner NGO
The international partner NGO is an organization with over 50 years of practice in rural development. The NGO is engaged in all major agricultural commodity chains—reaching from cotton, tea, cocoa and coffee over livestock and diary to soy, palm oil and sugar cane. Besides, it is also involved in the gold and textile supply chains. The NGO is active all over the globe and in West Africa runs programs in Cote d'Ivoire, Ghana, Liberia, Nigeria and Sierra Leone.

The NGO's special fields of action is on public-private partnership arrangements and to bring different stakeholders together. Its engagement in UTZ capacity building in Ghana's cocoa sector started around 2012. It first established the link between interested TNCs and leading LBCs in the sector

The partnerships approach is the strength of the organization. Besides the described fields of action, the partner NGO is present at all important sector platforms and plays the role of an important stakeholder-connector in the sector. During the interviews with staff members, some self-descriptions of the NGO were given: Interviewees saw the partner NGO as an "expertise center" (interview with several staff members of the partner NGO in April 2015) which they rather understood as "a social oriented company that seeks to work with the private sector to drive development. But narrowly NGOs work more with civil society…" (Interview with several staff members of the partner NGO in April 2015)

What becomes clear from these statements is that the organization is not to be understood as a civil society representative in the sector, even if it is often depicted as such. It is an organization which follows a clearly set agenda of market-based solutions to development and which strongly advocates for this approach in its role as NGO, representing the civil society section in the several above described platforms. Not only does the vision and mission statements on the websites clearly indicate neoliberal views, but also the local staff members' statements repeatedly make recourse to market-based solutions to development as for instance with the following:

> "You know the whole thing is, yes we see poor farmers, poor people over there, low income and poor communities being people who need support, but they need to be supported in a manner that will fit the market arrangement. I mean it is not always that the market is right but at least it drives action. And every action that you take it needs to be consistent with the market realities. So, yes, the market says 'we want sustainably produced cocoa.' How do we get sustainably produced cocoa? How do the arrangements need to be put in place to ensure that this is transmitted from the farm and the farm gate to the consumer by the various actors in-between. So, definitively, whether we like it or not, the farmer is key, the farmer is important, but the farmer needs to be understood so that the market takes advantage of that kind of understanding." (Interview with several staff members of the partner NGO in April 2015)

For some of the interviewed staff members, the strong focus on markets and supply chains as means to foster growth seemed to outweigh the perspective of the individuals which are behind the production base. While the following statement surely does not represent the official perspective of the organization, it gives some insights on how interventions are approached by the implementors who are active in the field.

> "People often today talk about youth in cocoa. What is COCOBOD doing about youth in cocoa? What serious measures is COCOBOD putting in place to ensure that we can get the youth into cocoa and can get the older people in the cocoa industry out so that more and more young people can go into cocoa and do it better?" (Interview with several staff members of the partner NGO in April 2015)

During the interviews with the NGO's staff members, no concept what should happen to the elder farmers which are hoped to be replaced by younger, more efficient ones, has been provided.

The partnering LBC

As already carved out in Chapter 5, LBCs in Ghana's cocoa sector are directly dependent on a constant cocoa supply by farmers. Both, that is cocoa production sustainability and a stable or increasing market share, are hence the two most important interests of any LBC. Thus, the interests are basically congruent with the ones of the project's lead firm. The main difference is, that for the partnering LBC, the good relations with the farmers are crucial whereas this is not a primer interest for the project lead firm, who finally buys the beans from COCOBOD. For the partnering LBC, its participation in the studied project brought several advantages which supported the fast increase in the company's market shares from 2012 onwards till its drop-out. One, certification pulls farmers to an LBC because it comes with several advantages that other LBCs do not provide, and two, certification increases farmers' outputs and sustainability. The following statement by a representative of the partnering LBC underlines this trend which will be further explored in this chapter:

> "Certification is basically a method in order to achieve sustainability. So the main purpose of certification was to encourage farmers to stay on their farms and to keep on producing the cocoa, because of looking at the challenges or the fear that farmers might migrate into other businesses or migrating out of those cocoa growing areas to cities to get a better living." (Staff of the partnering LBC in April 2015)

The view on farmers as business instruments by this interview partner is very similar to the one articulated by the project lead firm's representative. Certification directly coincides with the intrinsic interests of the company. In 2015, the partnering LBC's target was to achieve 100% purchase of certified beans. But the business is risky and needs a lot of investment. The cooperation between the partners of the studied project therefore was a constellation that brought some security to the company:

> "So they [the project lead firm, author's note] have a market for UTZ cocoa. And in certification it is a risky business, if you do it on your own and you don't have a final buyer, because you invest a lot of money and you don't get a revenue. So in our case we have guarantee, we have secured the buyer who are our partners and then it is our partners who decide which standards they need." (Staff of the partnering LBC in April 2015)

Nevertheless, the findings which will be presented in Section 7.2 indicate that this security has a downside, too.

The partnering LBC was the main implementor of the project's activities and its extended local business network was used to establish the Internal Management System needed to run the project.

7.1.2 The Implementation Process

"They said they will bring some people here to train us on how to do the cocoa well. So when we ferment it for 6 or 7 days then it will be certified. So they told me to register 25 of my farmers. So I looked at the responsible ones who sell their cocoa only to me and I registered them. After I sent them then they brought an officer here to show us how to prune and they made me select a lead farmer to be trained for two weeks. When he returned he came to teach us what he learned that we have to prune the trees and remove mistletoes and so on. We should also plant trees to give the cocoa shade and to space the cocoa so that it doesn't get crowded. We should also not wait too long to crack the pods when we harvest the cocoa. When we crack it and store it we should turn it out in three days and then on the sixth day we will dry it. It will become a chocolate color and then we can call it certified. They said they will bring us a label and if all the 25 of us are able to do what they have said then they will stick the label on our cocoa as certified. Then they will bring us a premium." (PC of the partnering LBC in Body type 2 in July 2015)

This description of what the studied project's staff told the interview partner before the start of the project provides a telling summary on the main drive and local perceptions of the project. First, the way how the interviewed PC describes how he received the message of the project tells the reader about his position in the studied project. Using expressions such as "they told me to register...", "we have to prune...", "we should do..." reveals that the interview partner rather experienced the project as something he has to adhere as in contrast to something to what he can actively contribute. It seems that the participation level in the process even of local key persons such as PCs apparently was low. Secondly, a focus on product qualities to be achieved through particular post-harvest treatments, at least in the perception of the interview partner, becomes apparent. Thirdly, there is a special role of the premium for the implementation of the project.

The aspects mentioned here are all important parts of the three main fields of action during the implementation process of the project which have been identified during the study: (a) organizing farmers, (b) diffusing the particular interest of achieving long-term supply with high quality beans, and (c) increasing control. While the level of farmers' participation is an aspect of the mediating action field (a), the focus on product qualities is an ingredient of (b), and the premium plays a key role for (a) and (c). But before these mediating actions could be conducted, project partners had to set up an enabling structure which allowed the implementation of the project. This structure was the Internal Management System (IMS). In the following, first the IMS is described, which allows a better understanding of all activities within the project. Subsequently, the three main fields of action during the implementation process of of the studie project will be described and analyzed.

IMS as superordinate, enabling structure
The studied project started in 2012 and in 2015 already 15,000 farmers throughout the Western North and parts of the Ashanti Region were UTZ certified in the course of the project. The project operated through a highly sophisticated and decentralized IMS (till 2014 still called Internal *Control* System), which was part of the UTZ requirements. The IMS was built upon the already existing marketing structure of the partnering LBC in 15 of its operational districts. The districts were clustered into three project zones, each consisting of five operational districts and their numerous communities. There were IMS staff members at every respective level of the hierarchical system. The leading body of the project was the steering committee. Consisting of representatives from the three project partners, namely the sustainability managers of the project lead company and the partnering LBC as well as the project lead officer from the partner NGO, the steering committee oversaw the progress of the project and took all major decisions. The IMS covered the two main parts of the project: extension delivery and traceability. Whereas the extension system was completely newly established and all positions were new in the field, the traceability system became part of the existing marketing channels of the partnering LBC. All IMS staff members were working under the LBC. The following Figure 7.1 illustrates positions and responsibilities in the IMS.

Directly under the steering committee, the project officer was in charge of coordinating all activities in the three zones. At the zonal level, zonal managers were responsible for the coordination of all activities in their respective zones and had to send monthly evaluation reports to the project officer. Zonal managers oversaw the work of the IMS officers who were in charge of the direct implementation of all certification issues in one operational district. The IMS officers had to ensure that all internal requirements were met by the participating farmers and worked closely together with the extension agents of the project. While the extension agents taught farmers with the general GAPs, the IMS officers conducted trainings on all the requirements of the UTZ standard which go beyond the general GAP training. In the studied cases, the IMS officers were in charge of all

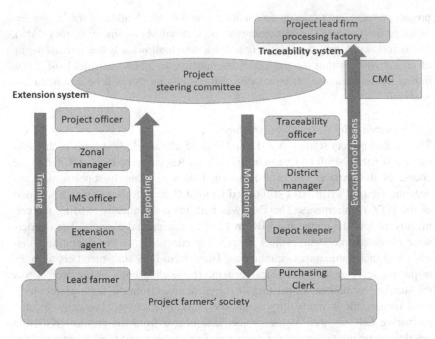

Figure 7.1 Internal Management System of the studied project. (Source: own elaboration based on interview findings)

IMS documentation duties at the district operational level, too. The project extension agents were hired through COCOBOD for the partner NGO and therefore were located at the district CHED offices and served as a connector between public and private extension systems. They received the same training as COCOBOD extension agents and, next to the GAPs training, were responsible for technical assistance of the farmers targeted by the studied project. CHED district managers were aware of the work plan of the studied project's extension agents and tended to accordingly allocate COCOBOD agents less to the communities where the agents of the studied project operated. The project's extension agents circulated around a huge number of project communities. Hence, most of the time, the training of the society members was conducted by the lead farmers of the group and the extension agent and IMS officer only came in once in a while. Lead farmers were not contract workers but as per definition of the UTZ standard carried out all their activities on a voluntary basis as local resource persons. Besides conducting the two-weekly training, lead farmers were also in charge of internal inspections and record keeping of farmers' program activities. The farmers had to organize themselves in a producer society which was supposed to have a number of key positions, next to the lead farmer such as treasures, organizers or a social welfare officer. At the traceability part of the project, the PCs, contracted on a commission basis for the project LBC, were the first persons who did the quality check and segregation of the beans. They had to record the number of certified bags the farmers supplied to the project and were also in charge of the delivery of several input items. Beans were kept separately along transport and storing in the local depots. The district manager, who oversaw all PCs in the district and the depot keepers, had to assure that all UTZ certified bags remained separated along the marketing chain and had to keep record on all district purchases of conventional and certified beans. There was a traceability officer at the top level of the marketing chain who monitored activities in all three project zones. The certified beans were transported from the local depots to a designated CMC warehouse, contrary to the conventional beans which go to random CMC warehouses. The project lead firm had an agreement with CMC on the amount of certified beans that it wanted to buy from from the partnering LBC and on the separate handling of them. Finally, UTZ certified beans were submitted to the project lead firm's factory in Ghana or shipped to one of its factories abroad.

Fields of action

a) **Organizing farmers**

After setting up the IMS structure, communities with a sufficiently high number of interested farmers had to be identified. Sensitization activities were first conducted by the IMS team (zonal manger and IMS officers), sometimes accompanied by sustainability staff of the project lead firm, in those communities where the project LBC had a high record of purchase and where a larger group of farmers clustered already around its PC. In the initial meetings, farmers have been informed about the benefits and requirements of their participation. Based on the interview statements, it is very likely that in many cases, the reason given to the farmers to participate in the project was the need to improve production in order to remain competitive as it is exemplified in the following statements:

> "He also said he has noticed that our yield of cocoa beans has reduced drastically and that they have a group that only deals in cocoa and they want to be in collaboration with [the project LBC] so they can teach us and show us how to get quality cocoa that is standard for the market or the way the group likes it." (Lead farmer in Juabeso type 1, June 2015)

"What they said to us was that Ghana was losing their quality and that they want to get the quality back." (PC in Juabeso type 1, June 2015)

"Then he said that we should follow the instructions they give us about the cocoa oth-erwise there will come a time that when your cocoa has not been certified, no one will buy it." (PC in Bodi type 2, July 2015)

Especially the last citation shows that some farmers received a message that uses a certain degree of pressure for joining the project.

As the reputation in Ghana says, cocoa farmers are rather skeptical towards new things. Indeed, many interview partners described that initially not many farmers were interested in joining the project. PCs mainly used their personal ties to create the group out of their most loyal farmers. That is why in many cases, many members of an extended family are part of the society. In all interviewed communities except one, the PC appointed the lead farmer, as illustrated by the following statement:

"So when the PC told me that he wanted me to do that job I did not refuse." (Lead farmer in Bodi type 2, July 2015.)

During the initial period of the project, the IMS staff had difficulties to gather enough farmers for the project and therefore, during the early phase, there were no special conditions for participation. After the three years of project conduct from 2012 to 2015, the attractiveness of the project had drastically increased. Many farmers became interested and wanted to join societies in their communi-ties, even if prior they used to sell to other LBCs. After some time, the registration for participation had to be closed because maximum participant numbers of around 50 members per society had been reached. The limitation of member-ship was a direct result of the market-based approach of the project, where the supply is guided by market demand. That means, if too many farmers would become certified, processors would end up selling certified chocolate products as conventional ones and lose the investment in certification they made. Due to the limitation in membership, some farmers attended the training sessions even if they were not officially registered in the project's society.

"We started as a very small group but now many people have joined. Some have even not been registered but I have been registered for a long time now. I have my card so I'm part of everything. When they realized things are going well, then they all came to join." (Society member in Bodi type 2, July 2015)

While a representative of the studied project saw the training and the prospect of production increase as the main reasons for the rapid spread of farmers' interest in the project, in almost all interviews, farmers mentioned the premium as the major incentive for joining, followed by the distribution of inputs such as Personal Protection Equipment (PPE) such as overalls, boots, hats, goggles, and chemical inputs.

> "The giving of the overall and the other equipment was what made them come and join. Especially the bonus. It really pulled people into the group." (Society member in Bodi type 2, July 2015)

> "We were very happy because we knew we were going to get chemicals from them so we were happy." (Chief farmer in Juabeso type 1, June 2015)

> "They told us that if we produce the certified cocoa and we sell to them they will also bring us some equipment and also some premium. That's what they promised us." (PC in Bodi type 2, July 2015)

In a setting of very difficult access to any additional equipment and inputs and lack of finances to buying them, the delivery of these small materials is a very simple but efficient tool to trigger interest among farmers. The premium, by then as low as 8 GHS[1] per bag of 64 kg standard compliant cocoa beans, was still the most often mentioned incentive for farmers to join. For people living in poverty, even a small amount of additional cash constitutes a relevant additional income and hence is an incentive to participate.

While the access to inputs was said to be what motivated the farmers to be part of the project, reports on the actual facilitation of the access to chemicals are not conclusive. Some interviewees reported that there were some plans communicated, that Syngenta would bring some products to farmers which would later be deducted from their sales revenues by the PC. However, at the time of the interviews in 2015, this had not happened yet and many interviewed farmers were disappointed by it.

Looking at the described aspects of the organization of the farmers for their participation in the project, the concern that the farmers are only marginally involved in the whole process and only have the role of passive recipients of project requirements is further increased. The descriptions of the members give reason to assume that they participated mainly because of the small material incentives and that they did not have a clear overview of the overall context of the project and the meaning of their participation. Key positions were usually

---

[1] In 2015 an equivalent to 2 Euro.

assigned by the PC without any internal decision being made by the group, as it is usually the case in self-organized groups. All in all, there is reason to be concerned that the emancipatory content for farmers in this constellation is rather low and that the self-organization implemented by a top down approach is mainly instrumental for the functioning of the project as in contrast to an intrinsic value of self-organization.

b) **Diffusing particular interests**

Analyzing transnational CSR in the context of the consent and control framework, it is the enforcement of the project lead firm's particular interests which is of main interest here. Already the process of pooling forces with the two project partners is an important act of building an alliance which has even broader capacities to diffuse the main interests. In the given case, all three actors share the same interests of keeping cocoa farmers in their position, or at least as many cocoa farmers as needed to achieve an effective production increase and a lasting cocoa supply. Interventions therefore have to tackle problems which are regarded as risks for the sector. As described above, one, is the situation of extreme poverty of many cocoa farmers and the weak infrastructure in many communities. The second is about farmers' farming capacities which are regarded as low. While the studied project is not efficiently responding to problem one—besides the delivery of the premium, it is mainly dedicated to improve problem two, which is regarded to also contribute to overcome poverty of cocoa farmers.

Accordingly, the main activity in the course of the studied project was the training of society members with the UTZ Code of Conduct. In the first year, training took about six months and the first audit took place after one year. The UTZ code is based on a continuous improvement approach so that every year, over four years, a number of additional topics have to be treated. The whole extension system of the studied project was based on a 'train the trainer' approach. The Code of Conduct and how to train the farmers were taught to the whole IMS staff. It was the UTZ country representative who gave training to the top level staff of the studied project and then mainly the partner NGO was responsible for training the project staff. Zonal managers and IMS officers received several week-long workshops.

"Yeah, we were trained on the Code of Conduct, we were trained on how to facilitate training, how to communicate well to the farmers, so that they understand well, your gestures, so that it might not be like you are insulting them, how you make your face, how you will turn of your hands, like in our local culture..." (Staff of the studied project, July 2015)

IMS staff members, in turn, were then mainly in charge of training all lead farmers of the societies in several week-long workshops and additional meetings. In some cases, sustainability managers from both the project lead firm and the partnering LBC also led the lead farmer trainings themselves, as for instance the following statement informs:

> "In the beginning it was Mr. [name of zonal manager] that started the training with us. He was the one we were learning from. It was later that heads of the Company also started training us. Teaching us a lot of things and we even had to take pen and books along in order to jot a few things down. During the training session, questions are asked and lead farmers are required to answer. We were given sheets of paper with questions on them and we answered them. The papers were collected and taken away." (Lead farmer in Wiawso type 3, July 2015)

It is an interesting aspect that lead farmers have to pass a kind of exam and apparently were surprised about it. This also feeds into under a) discussed lack of participation and indicates that farmers rather hold an "object-position" in the process.

Lead farmer trainings comprised the theoretical parts on the internal standards and Code of Conduct and practical exercises on the application of chemicals and GAPs. Moreover, lead farmers were taught how to conduct internal inspections for the pre-audit, how to fill the farmers' passbooks and how to keep record of the training activities. After the successful participation in all specialized trainings, lead farmers started to pass their knowledge to the society members. The societies met every two weeks for training either on a demonstration farm to conduct a Farmer Field School (FFS) or around the PC's depot to learn the theoretical parts of the requirements with the UTZ learning material. The Table 7.1 shows the training contents as required by the UTZ Code of Conduct.

UTZ certified, before its merger with Rainforest Alliance, described the goal of its Code of Conduct as to offer expert guidance on better farming methods, working conditions and care for nature. This in turn would lead to better production, a better environment and a better life for everyone, the organization promoted its program (UTZ—Rainforest Alliance 2020a). Looking at the individual requirements listed in Table 7.1, it seems that it is very difficult to implement all of them. Some requirements, as for instance the farm management requirement "ensuring access to inputs", the farm practice requirement "diversification of production" or the social and living conditions requirements "access to education

Table 7.1 UTZ Certified Code of Conduct and Cocoa Module Version 2015

Farm Management	Farming Practices
Measures to optimize the yield	Choice of suitable planting variety
An Internal Management System for groups, with responsibilities including:	Farm maintenance
	Soil fertility management
Transparency on the premium and how it is divided	Diversification of production, to support ecological diversity and economic resilience
Ensuring society members have access to inputs such as training and materials	Integrated pest management
Arranging annual internal inspections	Responsible and appropriate choice and use of pesticides and fertilizers, and records of application
Record keeping	Irrigation
Risk assessments	Product handling during and after the harvest
Training and awareness raising	
Recording of volumes in the UTZ Certified traceability system	
Social and Living Conditions	**Environment**
Application of national laws and ILO conventions regarding wages and working hours, including the living wage concept for individual farms	Efficient use of water and energy
	Waste management
No forced labor or child labor	Promotion of ecological diversity
Freedom of association and collective bargaining	Protection of nature
Safe and healthy working conditions, including:	No deforestation of primary forests
Protective clothing for work with chemicals	Respect for protected areas
Safety training of workers in their own language	Protection of endangered species
Gender equality	Reduction and prevention of soil erosion
No discrimination	Measures to adapt to climate change
Freedom of cultural expression	
Access to education for children	
Access to decent housing, clean drinking water and health care for workers and their families	

(continued)

Table 7.1 (continued)

Cocoa Module
Farm Maintenance
At least 12 shade trees per hectare
Access to shade trees seeds or seedlings ensured
Postharvest processing
Appropriate fermentation
Appropriate drying methods that prevent contamination
Drying to an appropriate moisture content
Preventing beans to get wet during storage and transport
Cocoa meets agreed quality standards

Sources: UTZ Certified Code of Conduct and Code of Conduct Cocoa Module 1.1–2015.

and access to health care" as well as the environment requirement "measures to adapt to climate change" are clearly difficult goals to be implemented in a framework of an individual project and are of rather structural scope. Moreover, some requirements tackle deep-rooted cultural behaviors and practices such as gender relations and discrimination against particular social groups. The question arises whether a project with a strong top-down nature as the studied project is the right mechanism to tackle such important aspects and will be discussed in Chapter 8.

However, findings indicate that during the trainings, the focus was on a number of practices which were considered more relevant for the local setting or are closer to the particular interests of the project lead firm and its allies. When being asked about the training contents, social and environmental topics were almost never mentioned by interviewees. The few social and working requirements that were mentioned are the wearing of PPE and other protective measures during chemical application as well as the ban of heavy work for children. Both are important aspects and would also present considerable risks to the project lead firm if problems with them would appear in the certified supply chain. Environmental training issues that were stated in the interviews are the correct disposal of plastic containers after chemical application, not to plant trees too close to a slope and not to apply chemicals on cocoa trees next to water bodies. By far, the training contents which were most often mentioned were the agricultural practices pruning, mistletoe removal and the correct application of chemicals as well as the postharvest practices of six days fermentation and appropriate drying. Since most of the farmers already did some pruning and mistletoe activities before the start of the project, the correct way of chemical application and the six days fermentation were the practices that have been most often mentioned as the new practices taught in the trainings.

> "Ok what they have taught us is that we should ferment the cocoa for six days and turn it twice and afterwards we dry it, so those who have joined we make sure that they do that and I also do same because if we don't do that and break it they will find purple bean in it and it will not be accepted. And we are doing this to maintain the quality and so how they have taught us is what we are doing." (PC in Bodi type 2, July 2015)

> "Before the program we didn't know how to measure the chemicals so we were using a lot but since the program started they have taught us how to measure so now we use less." (Society member in Akontombra type 1, July 2015)

While new farming practices can indeed show a positive return to farmers due to their increased productivity, the additional work load in post-harvesting activities and the thereby achieved improvements in beans' quality is only remunerated with the premium. It is the project lead firm who might receive a higher price for their high-quality cocoa products.

c) Increasing control

The project uses a number of tools which allow for a high level of monitoring and sanctioning, hence control over farmers' behavior and project implementation. Key instruments were the use of contract and registration forms, the obligation to keep record on all conducted activities, as well as farm inspections, audits, and a traceability scheme for internal marketing.

In order to become a member of a project society, farmers had to register and sign a contract. In the registration sheet, information on farmer and farm were collected, as for instance the number of farm plots, production area size, number of employees, educational level of the farmer. In the contract form, rights and duties of both sides, that is the farmer and the project were explained. The breach of the contract could lead to the exclusion from the project and any benefits. While most interviewed farmers mainly remembered their duties to apply GAPs and to sell to the LBC, the complexity of the contract is described with the following statement of an IMS staff member:

> "After we register them, we explain to them what the registration form talks about, we have a contract form which I think is also mainly three principles here: Obligation of the company, obligation of the producer that he must adhere to the standards, that is what we train about, to do a continuous training and to apply any technical recommendation. And then give all their agric information to the internal inspector and to allow their access to all his or her production units, that is their farm, and to any documentation that they were doing. To accept the internal standards and external sanctions and to put the corrective measures into practice if any. You have to accept it, you don't have to do like you are not accepting it because the law binds you here. So therefore, you have to understand to report any change or variation of condition of production at your farm if anything you have to tell us. So right now, you have joined the group, so if you make a mistake, the whole group suffers. And then one thing is that you deliver your beans to the company. That is one." (IMS staff, May 2015)

It is instructive that the interviewed IMS staff member only describes the duties of the farmers while not telling any details about the implementors' obligations. It also becomes clear that membership means a tight jacket of requirements and that farmers here again are seen only as passive recipients who have to align themselves with the project demands.

In almost all interviews, the society members could not make any difference between registration and contract and had forgotten about the content details.

The fact that many of them are illiterate and only hear the content during the first meeting when it was read to them apparently reinforced that gap. But, nonetheless, the arrangement seemed to work out and participants respected the three main aspects that were to attend the meetings, to follow the Code of Conduct and to sell their beans to the project LBC, the "three basic laws", as project staff called them.

Society members received passbooks in which they were supposed to record all their farming and training activities as well as their beans sales. In most cases, it was the lead farmer who filled in the training and farming details for the society members, not only because many of them are illiterates but also to assure the quality of the records. Regarding the certified sales, it was the PC who noted the quantities in the society members' passbooks. The detailed record keeping on all conducted activities was done at every level of the IMS hierarchy. The lower hierarchy staff had to report the monthly activities to their superiors. At the district and zonal levels, there should be documentation officers but in the investigated cases there were no such and the interviewees stressed the huge workload coming from documentation. This is also the case for the lead farmers and the PCs but contrary to the IMS staff, they were not remunerated for these activities since they were not regarded as system employees but volunteers.

Possessing data on farm sizes and cultivated area is an important asset to be in the position to forecast productivity of farmers in the given region, and hence a key business and sustainability tool. GPS farm data collection was already conducted in 2015 but was in its initial stage. Attempts to strengthen farm data collection were reported by project staff. The extension agents of the studied project were trained to measure farms by COCOBOD which also seeks to improve the national data basis on farm sizes. Based on the farm data, UTZ supplying companies are able to forecast estimates of UTZ certified yields and to monitor environmental practices. The tool therefore became a central approach in the program. In 2019, the Rainforest Alliance/UTZ merger launched its Cocoa Assurance Plan. Requiring stricter GPS mapping for all certified groups is one of the main approaches of the plan (Rainforest Alliance 2019).

All UTZ certification projects needed to be assessed independently by third-party certification bodies. But before external auditors were called, a main practice was an internal inspection, also called pre-audit, to assess whether farmers are ready to go through a real audit and get certified. Such an internal inspection usually took place for the first time approximately one year after project start, when the IMS officer estimates the group getting ready for external audit. Pre- and external audits were conducted every subsequent year. Each participating farm had to be inspected for the pre-audit while for the external auditing only some farms were selected. The following statement exemplifies how a lead farmer perceived the act of inspection.

"When we do train the farmers, maybe in 6 or 7 months' time we have an internal inspection team and they come and inspect the farmers farm that what you are learning, they using or not. Because before you can be certified you need to do pruning, removal of mistletoes, weeding, and good agricultural practices. So the internal officers will come to your farm to see if something you have to do and you did not, then they report or they give you a mandate. So before everything will be okay then they recommend you." (Lead farmer in Akontombra type 1, July 2015)

The mechanism of control becomes very clear in this statement. Each individual farm was assessed whether it had successfully applied the techniques. Only if this was the case, the farmer could pass on to the next level to finally become certified and receive the premium. During this internal inspection, the lead farmers who mostly conducted the inspection used the internal inspection checklist. The English document contained over 40 control points in the fields of farm establishment and rehabilitation, soil and fertility management, pesticides and fertilizer application method and equipment, empty pesticide containers, pest and disease management, pesticides handling, environment, protection of nature, harvest and post-harvest as well as worker rights—all aspects to be assessed by the lead farmer for up to 50 society members.[2] This is likely to be a highly time consuming and challenging activity for the voluntary working lead farmers.

Farmers knew that all of them had to comply else the whole group was not able to pass. Besides this kind of group pressure, the premium functioned as the main tool to threaten consequences of non-compliance, a sanctioning mechanism that has been mentioned in many interviews:

"It was during the training that we were told that if we do not qualify after inspection we won't get the bonus. We were also made aware that when we submit our beans to the headquarters and it is detected from the machines that there are chemicals in them. We are supposed to pluck the pods after three weeks from spraying. If they are plucked from the three weeks duration it will be detected by machine at the headquarters. If this should happen you will be denied bonus. So we were cautioned to follow the instructions to the latter." (Lead farmer in Juabeso type 1, June 2015)

"They told us in the beginning that if you don't dry your cocoa well off, if the PC comes to inspect it and it wasn't fermented for the right number of days then you could lose your premium so everyone is happy to do it well so that the small premium we are getting from this will not be lost." (Society member in Wiawso type 3, July 2015)

---

[2] A copy of the checklist is attached in the appendix 7.

If all farms were estimated to show a good degree of implementation, the IMS officer communicated the readiness to the steering committee and the real audit was arranged. It was external auditors from the contracted third-party certification bodies who came to enter and inspect a number of selected farms. The procedure did not happen unexpectedly as the statement of one lead farmer indicates:

> "So when we get to know that they will be coming around then I have to go and check the individuals if they have really done what they were taught so that when the officials come around they will know we are doing our job right. So we pick the weak ones and go to that person's farm and do thorough check on the farm so that when the officials come around they will know we are doing the right thing." (Lead farmer in Juabeso type 2, June 2015)

This statement by itself tells about the contradictions of the practices of independent auditing. Prior to the act of auditing, farms are prepared to look nicely so they can easily pass. Nevertheless, from the control and consent-perspective, the pressure to follow the new rules which makes farmers align their practices with the demands becomes clear too.

Once an audit was successful the information was passed on to UTZ and the certificate issued. It is the project LBC who held it being the representative of the societies. The premium was distributed to the farmers according to the number of bags they had delivered.

During the internal marketing chain, a traceability system was applied. UTZ certified cocoa bags, in addition to the COCOBOD tracing numbers, were sealed so that they could be traced back to the society of origin. In the frame of the studied project, the PCs checked the quality of the beans before they recorded and bought them as UTZ certified beans from the society member. After this, they kept them separated from conventional ones in their depots. The bags were segregated along the whole marketing chain as described above till they are delivered to a designated warehouse at one of the ports where a regular quality control by QQC took place. Subsequently, certified beans were either delivered to the project lead firm's factory or shipped to one of its plants abroad. While traceability was assured along the national chain in Ghana, UTZ allowed the Mass Balance System, meaning that in the manufacturing process, certified beans could be mixed with conventional ones according to a determined percentage.

The given descriptions make it possible to reconstruct and understand how a set of new mechanisms of monitoring through different means of data collection, and of sanctioning non-conformance by the exclusion from the project and the loss of the premium was established in the course of the studied project.

7.2   Project Experiences from Different Perspectives

7.2.1   The Experiences of the Implementers

As described in Section 7.1, the implementing partners of the studied project cooperate closely and became a strong joint actor for the purpose of project implementation. Representatives from the three project partners share the same view on the rationale for the implementation of the project which is the need for production increase and training in business competences in order to overcome farmers' poverty. But besides this common understanding, depending from their different positions and responsibilities in the implementation, their experiences and concerns differ strongly. While the project lead firm and the partnering NGO's staff members pay more attention to the organization of the sector and the role COCOBOD should play in it, the partnering LBC's representatives are concerned with the challenges during the implementation process and risks for the future of the company.

Staff members of the project lead firm were less directly involved in the implementation on the ground and did have less to struggle with the requirements of the IMS; they rather read it from the IMS staff reports. In the interviews, they presented the project as in an advertisement, highlighting it as a full success story where farmers benefitted from the training, especially because of the business component of the training that teaches them how to save their money. Nothing was mentioned about problems during the implementation phase at the local level, but where more emphasis was placed was the overall functioning of the cocoa sector in Ghana. According to one representative of the project lead firm, for a well-functioning cocoa sector in Ghana, the involvement of COCOBOD is a positive thing but it should reduce its competences and focus on regulative measures. Important services in the local sector should be only carried out by the private sector and COCOBOD should coordinate and monitor activities in order to avoid duplication of efforts. In general, the interviewee stated, the cooperation with COCOBOD had substantially improved in the course of the project:

"In an exponential manner it is getting better and better. Initially COCOBOD of course was having a lot of governmental bureaucracy with government changes and all that. And you see that there is some level of resistance within the system but these days things are getting better, people are appreciating things better and the collaboration is getting far, far better." (Representative of the project lead firm, May 2015)

A good relation with the board seems to be of importance for the interviewee. Besides, the use of the strong word "resistance" indicates the perception of the interviewee that COCOBOD is confronted with dynamics of change which might not be completely in line with the board's interests.

The interviewed staff members of the partner NGO share these views on the importance of COCOBOD and also emphasize the continuous improvement in cooperation with COCOBOD. Here too, the wish that the public agency should assume a stronger coordinative role and move away from direct involvement in the supply chain is articulated repeatedly. It becomes clear once more, the two entities share a similar idea of neoliberal sector organization.

Knowing the challenges at the production base better than the interviewed staff members from the project lead firm, interviewed representatives from the partner NGO see the strength of certification in providing improved extension services and inputs to the farmers. Ghanaian staff members of the NGO somehow hold the position of bridge builders between the NGO from the Global North and locals in Ghana. One interview partner interestingly pointed to the problem that Western supply chain stakeholders should understand the difference between Ghanaian cocoa farmers and Western farmers and see the challenges of implementing complex requirements:

> "But the fact of the matter is that these people [standard setters, author's note] have not experienced the reality of a farmer and coming from that far, they think that it is very easy to get the farmers to do all these things". (Staff of the partner NGO, May 2015)

In the course of the interview, the interviewee seems to be in between the discourses of theory and practice. On the one hand, he sticks to the idea that certification is the key tool to achieve important sector goals as for instance to get farmers an additional income through the premium, to get farmers organized and to reduce child labor. On the other hand, he raises concern that besides successful certification going on, farmers still live in poverty. The main reason for that, he argues, lies in the level of productivity which still has to rise significantly if farmers are to overcome poverty. Generally, the interviewed staff members of the partner NGO emphasize broader local sector transformation.

During the interviews, the explanations for farmers' challenges and local sector functioning given by the interviewed staff members of the partner NGO in 2015 were closer to the local circumstances than the ones stated by the project lead firm's representatives. But it was the interview partners from the partnering LBC who had the broadest knowledge on local conditions and most experiences with implementation. In the interviews, these staff members equally emphasize the positive effects of certification on farmers who were described as becoming more business oriented and empowered through the training. At the same time, compared to the other two project partners, interview partners from the partnering LBC have a much clearer picture of the challenges during the implementation process and how farmers reacted to the project. The major challenges which arose for the IMS staff and the whole company as they explained it can be grouped into two fields: one, they have to do with the high workload during the implementation process, and two, with the risk of losing farmers. Since the project was based on LBC's marketing relations, it was also these relations which are directly exposed to damage in case of failures. There was no similar risk involved for the other two project partners.

Regarding the high workload for IMS staff members, interviewees gave several explanations. For instance, they stressed the difficulties that arose due to the low educational level of some farmers which would bring the need for a lot of followup and coaching activities from IMS officers. Even the educational level of some lead farmers would be so low that they would struggle with the conduct of the internal inspections, as the following citation summarizes:

> "That is an issue we have…because in some communities they don't have an elite. All the elite guys have gone to school, secondary school they are all in the cities and there are some guys who have had the basic education but they are not that good. Some of them are even the community's secretary so you can imagine the kind of community we have. So what we do is that as they go round and they do their internal inspection sometimes we go back with them or we sit down with them and talk about the results. Maybe this farm you measured it, you remember. And they can tell you 'when I went to this farm, I saw this, I saw that, but I didn't know where to put it' [in the internal inspection checklist, author's note] and then you teach him 'it is this point where you have to put it. Sometimes too, they have a book, I know one man who has a book, and everything he sees he writes it in the local language and then he submits it to us." (IMS staff member, July 2015)

This explanation provides a good insight in the ground level activities of IMS staff members which have to make sure that the whole project approach works out. It equally shows that the level of knowledge about farmers and their communities for LBC staff is very high compared to the other two project partners.

Next to the high workload linked to the need for IMS staff at the local level to continuously follow up with farmers' improvement, IMS staff member described a huge amount of workload stemming from the documentation requirements of the program.

"It [the workload, author's note] is high! It is not easy. Because you are dealing with farmers. It is not easy working with humans. Some might understand you, some might not understand you. And apart from that when you need to get a documentation officer. […] The whole IMS system is documentation. It is documentation. When auditors come, you say you have done this – 'where is it?' You have to prove it! So we are combining field work with desk work a lot. So we should get a documentation officer plus we the IMS officers." (IMS staff member, July 2015)

This statement of this IMS staff member shows a degree of frustration of the interview partner. It seems that the double task of being continuously on the field while also being in charge of a lot of paper work which are both described as time consuming activities are difficult to combine and the lower level staff members of the IMS carried quite a big burden. The fact that documentation officers are officially required but not present indicates a malfunctioning of the system passed on the back of the most vulnerable staff.

The high workload has been stressed at all levels of interviewed IMS staff. Moreover, it has also been recognized for the PCs they work with. As noted above, PCs are in charge of additional product documentation. IMS officers are aware of the additional burden; one interviewee describes it as follows:

"But you should also keep in mind, they also have their work, some of them are farmers, some of them are traders, and they see that combining all these things is making their job very difficult." (IMS staff member, July 2015)

Only one PC complained openly about the increased workload and the voluntary approach:

"Yes so many additional work but they are not giving us the PCs anything, we are working just like that but nothing is coming out of it." (PC in Juabeso type 1, June 2015)

All other five interviewed PCs also mentioned the workload increase but did not stress the voluntary approach.

Next to the challenge of high workload, one IMS staff member also communicated the risk of failure he feels in his work and how a failure would affect the company.

"We have to tackle everything because of UTZ because the code is so strict and says 'do this' and if you don't do it, you will fail, and if you fail, your company fails, so you don't have to fail, you always have to be there. That is what we do. So we are there and enforcing it." (IMS staff member, July 2015)

This statement reveals a certain degree of pressure under which some IMS staff members seem to work. The fear that the whole company would get into trouble if IMS staff fails to correctly implement UTZ requirements and to get farmers through the process of certification become apparent here. This concern is closely linked to the sensitive relationship LBCs in Ghana have with farmers in the setting of the liberalized local marketing system without price competition (as described in Chapter 5). For the project LBC, the UTZ certification project was a strong tool for this competition. Being willing to join the project, many farmers abandoned their traditional LBCs and started to sell their produce to the project LBC. Still, the interviewees described the process as extremely risky, always facing the possibility to lose "their" farmers to other LBCs. It seems that certification is a strong tool of competition and helps those LBCs who are capable to implement the IMS to increase the number of farmers who sell to them. But at the same time, challenges are manifold and if not treated well can cause an adverse effect on the LBC. Most and foremost mentioned by IMS staff, is the fact that membership in participation is strictly limited. By then, the lead firm of the project, a cocoa beans processor, only needed a certain amount of beans which corresponded to the chocolate manufacturers' interest in UTZ certified beans. Hence, the number of needed certified beans determined the number of possible participants in the program. But since the whole marketing chain is based on estimates on how many beans certified Ghanaian farmers are able to produce, problems arose for the project LBC once farmers managed to produce more than expected and were not allowed to sell all of their harvest as certified. The following statement by one IMS staff member illustrates this challenge:

> "Rather we fuel some farmers, we cannot deliver as certified, they will be disappointed with us. That is a problem we are always facing. Because [...the project lead firm] will tell us, maybe 'I want to buy 1000', we look at our system, we look at our average, we do a rough estimate. If we do an estimate, then we assume that with these 20 hectares of land we say by Ghana standards we are supposed to get this, so here we can get roughly this. That is an estimate. And then if that doesn't work, we can do little about it. Unless [...the project lead firm] tells us 'oh, increase it', so we could buy more, but less is taken away by [...the project lead firm]." (IMS staff member, July 2015)

It looks like, even if the sustainability directors from both the project lead firm and the LBC affirmed the equality in the partnership and the joint nature of decision-making, finally, it was the marketing needs of the lead firm which directed how many farmers were allowed to participate in the project. The limitation of the need for certified supply caused two main types of challenges for the LBC: firstly, farmers might have produced more beans applying the standard requirements but are only allowed to sell the amount they have registered with—a fact that most of them are not aware of. Staff members of the project LBC report cases where farmers got annoyed that they only received the premium for a part of their beans and subsequently sold the rest to another LBC, meaning a loss of purchase for the company. Secondly, the PCs were continuously confronted with farmers who were eager to join the project but, in most communities, registration was closed and late comers were not able to enter the society.

> "Yes, that is a huge problem, it is a huge problem we are facing! Because you go to a group, and you can meet 50 farmers and you can realize that five farmers have not been registered, they are not in the project, but they still want to join." (IMS staff member, July 2015)

Hence, the project LBC being faced with these challenges at the community level advocated for an increase in the purchase amount of certified beans without prospect of success, since this was a challenge that is intrinsic in the market-based approach of certification:

> "It is a challenge we are always telling our partners to increase the tonnage for us to be able to get down this kind of problem. And then also the buyers have to pay more for their products. Yes, we understand the consumers also but sometimes it is difficult for us." (IMS staff member, July 2015)

Another reported challenge that puts the LBC at risk and that is similarly related to issues of external marketing is the delay in the payments of the premium. Farmers are often waiting intensely for their cash and every Cedi counts for them. If the premium is delayed, it is particularly the PC who has to justify it and who faces a lot of pressure.

> "So right now, if I don't have money to buy the premium cocoa I am worried. Because I don't want to lose my farmers and that is something that you need money to buy the cocoa. If a problem happens, all the work that I have done will be a waste." (IMS staff member, May 2015)

Yet, besides all challenges, the LBC's staff members see certification as an important tool to keep their strong position in the sector. Accordingly, the fear that other important LBCs such as PBC expand their engagement in certification was expressed and shows the competitive advantage for those LBCs which are able to implement the IMS. That this is only possible for bigger LBCs with a good funding base becomes clear with the following statement:

"Well, in those projects you have to rely on yourself and your own resources, you don't get any help from anybody. That is why a lot of LBCs are not able to do the certification. And that is why the public sector or the government cannot implement something like that. Because there is a lot of investment involved, a lot of monitoring involved, and a lot of…well…you need to be on top of things to be able to get certified and then to get a real and true certification, something, that is really impacting the farmers. A good impact needs a lot of work, a lot of monitoring, a lot of follow-up, a lot of back-stopping, and then you need to have the right people doing that. So, to be honest to you, when it comes to the project, we didn't have any support, its something you have to do on your own." (IMS staff member, April 2015)

7.2.2 Farmers' Perceptions on the Project and their Participation in it

In 2015, all interviewed society members had a very positive view on the project and expressed their happiness to be part of it. Following IMS staff explanations, this has not always been the case. When the project started in 2012, farmers were not interested in participation and rather skeptical towards it. But after three years of the project, the general attitude towards it had changed in the target-communities and most of the interviewed farmers who were not participating yet were eager to join. Farmers had seen that the premium was being delivered and PPE distributed, they had seen that participants were increasing their productivity; hence, the project became highly attractive for non-participants, too. In the interviews, the benefit by far most often attributed to the project participation by both participating and non-participating farmers is the premium. But the PPE and other equipment, particularly spraying machines, were mentioned often, too. Many farmers also describe the training itself and the resulting improvement of their farming output as a benefit. In a few cases, learning about health issues, as for instance how to protect and clean oneself appropriately after spraying was also mentioned as a benefit from the participation.

Next to the incentives for their participation, project members have been asked about changes in their farming practices and livelihood perspectives coming with their participation in the project, about their future perspectives regarding cocoa production, about their knowledge on the reason and overall functioning of the project, about their expectations of possible outcomes, and about the difficulties and challenges with it. In the following, the most important findings of farmers perceptions regarding these topics are presented.

Farming practices
The effect of their participation in the project that has been most often mentioned in the interviews is farmers' improvement of farming capacities. But such statements have to be analyzed cautiously, since dynamics of socially-desired answering might come to play here, especially because some farmers might have thought that the author was in some way connected to UTZ. The following statement exemplifies statements which seem to show some level of socially-desired answering:

> "When we started, we were following our father and we were doing the farming together with him. Things have changed from the olden days. It was very difficult to even plant the cocoa and harvest the beans in the olden days. But since we joined the group, they have taught us the process of planting the cocoa so as to harvest enough beans. But at first, we used to go through a lot of difficulties. The cocoa pods did not develop well and even plucking them from the stem was not easy at all. But since the group came and they taught us the measurement to use in the planting of the cocoa seeds, now it is very easy to harvest them when they are ripe." (Lead farmer in Juabeso type 1, June 2015)

This statement cannot be read out of context. Generally, the author got the impression that interview partners were afraid to mention any negative aspect of the project because they wanted it to continue and get the little improvements linked to it, particularly the premium. This is the most likely reason for such overly positive statements on the project which were predominant throughout the interviews. But in many other statements, which seem to be less intentional and more spontaneous, the improvement of farming practices and the increase in production due to the project are also highlighted, indicating that farmers really perceive this positive impact on their farming practices. The following expressions exemplify this general perception among participants:

> "Now that I follow their instructions this year has been good." (Society member in Wiawso type 3, July 2015)

> "At first my cocoa was getting rotten but now my cocoa doesn't rot." (Society member in Akontombra type 1, July 2015)

> "Yes. If at first I was getting about 6 bags now I get about 13." (Society member in Juabeso type 2, June 2015)

Most of the farmers describe the training as a positive thing that helps them to get their cocoa farming right. It is difficult to assess to which extent farmers really consider their practices before the training as different to what they are being trained on. But in the interviews, the dominant narrative was that farmers estimate the farming practices they applied before the project start as old and inefficient and that thanks to the project they were able to learn "good farming" what led their cocoa farming to improve, as the following statements illustrate:

"Before I joined the group, I didn't know how to prune the cocoa but now since I joined I know how to do that and it has helped my cocoa to grow." (Society member in Wiawso type 3, July 2015)

"When they teach us the things and you don't use it your cocoa will rot. They even showed us how to do the soil so that the cocoa will stand firm. If you don't do it your cocoa will rot and fall down." (Society member in Wiawso type 3, July 2015)

The collected data does not allow to tell if farmers rather repeat what project staff members have told them as the reason for the training or if they really perceive their former farming practices as less efficient than the ones taught in the course of the project. But there are indications that farmers really think this way, as for instance, when attributing a sense of modernity and progress to the project, as the following two answers to the question of the reasons for participation show:

"The reason is that I like to move forward." (Society member in Jueboso type 2, June 2015)

"They say "education has no end". So we have to keep learning because things are changing in the modern world and if we are still doing the old things…if you add knowledge to what you previously had it helps." (Society member in Bodi type 2, July 2015)

Above all, these statements represent the interviewees wish to further develop their farming skills and improve the livelihood situation which has been expressed in many interviews. It is noticeable that farmers are very keen to receive training if there is a possibility to do so.

Cocoa as a future perspective and other livelihood strategies
But these apparently perceived improvements in their agricultural capacities and cocoa production appear to not automatically translate into an increased attractivity of cocoa farming. Rather, interview statements seem to confirm Anyidoho et al.'s (2012) findings presented in Chapter 5. There is a clear trend that the

future in cocoa is mainly attractive for those farmers with very basic or no educational background as the following statements illustrate.

"For the farming, my dream is to make it big because it is a good business and if I do it well it will make me more profit. As I said I didn't go to school so this is the best I have to do." (Society member in Bodi type 2, July 2015)

"I can't leave the cocoa because I didn't go to school so that is all I have. It has never crossed my mind to leave the cocoa." (Society member in Wiawso type 3, July 2015)

In contrast, for many other farmers, cocoa is not regarded as a very attractive work. They highlight the physical struggle, the bad economic position and the missing respect which the community dwellers would show towards cocoa farmers as the following two statements show:

"We cocoa farmers have a lot of problems. In this town when you are a cocoa farmer you are not respected. Someone with a store or private business has more respect because their work goes on well. But we farmers every money we get goes back into the farm. Like the last harvest in January or February the money for that is in the bank. And every time you have to go and take some and spend everything. Any money you get you spend it on cocoa again. So the cocoa business is very hard. Very, very hard." (Society member in Wiawso type 3, July 2015)

"The other problem is that the farming is not even attractive. I have a friend abroad who says that the country where he is, farmers are the richest people so he doesn't understand why in Ghana farmers are poor." (Society member in Akontombra type 1, July 2015)

These two statements illustrate cocoa farmers' awareness of their vulnerable position in both, the GCCC and linked to their poverty, in the local society, too. In general, the future perspectives and whether cocoa is regarded as a livelihood strategy differs greatly among interviewed farmers. The views of the project society members were equally diverse as the views of the few interviewed cocoa farmers who were not participating in the project. While some see that cocoa is helping them in their life and do not imagine any alternative to it, others stress the difficulties to rely on it and envisage a diversified livelihood strategy:

"It is not possible to stop the cocoa business because it is what helps us and our children to strive forward, so we are going to continue for children to continue and those that follow. So is not possible to quit this job." (Society member in Juabeso type 2, June 2015)

"The way the cocoa is, it is seasonal. Some years are good and some are bad. So we just do it and we keep the profit so when things are bad then we use it to eat but to say that you will only do cocoa, it is very difficult. So I did something else. That's how I even got money to start. So if you rely on only cocoa you will suffer a lot. So even me when a new job comes I will do some." (Society member in Wiawso type 3, July 2015)

It becomes evident that there is a broad range of perceptions on livelihoods and the future role of cocoa among the interviewed members of the project society. This gives further indication that cocoa farmers are a heterogenous group with different endowments and aspirations. Still, livelihood goals which were expressed recurrently concern the interviewees' ability to save money in order to take care of their families, send their children to school and build their concrete houses.

"What am looking forward to in my life is that, once I am a cocoa farmer I should be able to save some money from the little I get so I can take my children far in their education so that when they are well off they will in turn take care of me. But as at now I have no help from anywhere unless myself, and I did not have anyone taking care of me so I don't have to mess up with the children but I have to help them so that they can have a good future." (Society member in Juabeso type 2, June 2015)

"What I want to happen in future is that I make a profit, am working and with every work you need to make a profit so my objective is that I will be able to... I don't even get to watch a television here, I don't have electricity and so many things that am lacking so my objective is to be able to care for my children and also to buy a plot in the city to build a house so that in future I can move there." (PC in Juabeso type 1, June 2015)

"What I want to in my life, my aim is to get the cocoa flourish for me so I can get property to cater for my children, for that's my aim that for my cocoa I will get some chemicals and knowledge to manage my cocoa so that I can get what am looking for to take care of my family." (Lead farmer in Bodi type 2, July 2015)

While these statements speak for themselves, it is worth notifying their common point in showing the drastic vulnerability in which many cocoa farmers live. What the interviewees state here are all wishes concerning simple basic needs[3] which are out of or difficult in reach for them.

This is of course one of the major reasons why for many interviewed cocoa farmers see the future of their children out of the sector, as they often stated with

---

[3] Next to the immediate absolute minimum basic needs for physical well-being (food and water), most of the existing basic needs approaches also acknowledge the human basic needs of shelter, clothing, sanitation, education, and healthcare.

a particular emphasis. Next to their poverty, another reason for that mentioned is the issue of land scarcity:

> "The land here is getting depleted so we have to work hard to send our children to school and further their schooling so that they can have a better future." (Society member in Wiawso type 3, July 2015)

> "Ok right now here and even in Ghana we do not have enough lands and the lands that I have my children are many and it won't be enough for them that's why I want them to go to school so they have good jobs in future so they can take care of themselves but no matter there will be one amongst them that will take over from to continue this cocoa farming." (Society member in Juabeso type 2, June 2015)

These statements indicate cocoa farmers' strategic thinking when it comes to securing the future for their families and achieving better livelihood situations.

Directly mentioned challenges
Interview partners were asked about both the main challenges with regard to the project and their general cocoa farming. Challenges regarding the project were almost not articulated in a direct way but became apparent only indirectly in the course of the interviews. It is possible that due to the above-described circumstances, interviewees were shy on articulating critiques on the project. At the same time, it is also plausible that the project participation does not bring immediate challenges which seem as strong as the ones which concern their overall livelihood situation, which are the following ones.

The lack of money is among the most often directly mentioned challenges. As already partly indicated in the upper citations, one of the major concerns is to have enough money to send children to school.

> "The thing that worries me is the children's schooling. The cocoa money is not enough for the schooling and also the cocoa work." (Society member in Wiawso type 3, July 2015)

A few farmers remarked that the premium would be too small and shily asked for it to be increased.

> "But the premium should be increased small. It will help us". (Society member in Akontombra type 1, July 2015)

There was no argument in the interviews that the regular cocoa price should be increased. This can be interpreted as an indicator of the low level of self-driven cocoa farmers' organization in the context of the annually fixed producer price by COCOBOD.

Other challenges directly mentioned include the lack or difficult access to chemical inputs and equipment like spraying machines. Besides, the weak community infrastructure is often stated and the absence of schools, boreholes or roads is expressed as a great challenge in their life.

Indirectly expressed challenges
In addition to these directly named challenges, some central problems related to the introduction of studied project crystallized in the course of the interviews and data analysis. These key issues include farmers' awareness on the content and the contract of the project, difficulties with some farming practices taught, social pressures and new patterns of inequality arising in the communities.

In general, there is a very low level of awareness and understanding of the content of both, the project itself and the contract which farmers entered. As briefly described in Section 7.2, the main understanding is that if the farmers correctly apply the requirements, the company gives them the premium and some other benefits. The following citations underline the missing awareness on the functioning and objectives of the overall project:

"When they registered us, they said that now when we get our cocoa we should bring it to them. If we don't then we will not receive our bonus. And also, when the supplies came some people didn't get some and they did not understand. This means that when the equipment's come, the officials look at how much cocoa you brought before they give you some of the supplies." (Society member in Wiawso type 3, July 2015)

"One was UTZ registration contract that if you dry your cocoa well and you pluck it and use the chemicals you've been asked to and not any other, they would give you one bag 8 cedis that was the contract." (Society member in Juabeso type 1, July 2015)

Such a low level of awareness on the reason for the establishment of the project reveals the small degree of farmers' active involvement in the whole implementation process. In a sustainability certification project implemented in the presented form, participating farmers seem to be treated as passive recipients which have to conform with the process instead of actively shaping it. This trend is further underlined by the tendency that society members' main focus is on their duties to fulfil the contract. No interviewed society member expressed that the company would also have a number of duties to fulfil, reflecting the description of the contract content cited in Section 7.1.2, where also only reference was given to the farmers' obligations in the contract. Besides, no one was able to tell the content of the contract and most of the interview partners, including PCs, did not even remember that they had signed such a document.

> "We did sign something during the registration but it was not a contract." (Society member in Akontombra type 1, July 2015)

> "They gave me some forms for them to fill with details like the size of your farm, the number of laborers you have in your farm, the number of people living in your house with you and you telephone number and so on. So that was the contract that they all signed." (PC in Bodi type 2, July 2015)

This missing awareness on the contract does not necessarily mean that farmers didn't sign or thumbprint two different sheets of paper, that is registration and contract forms. But what it clearly shows is the absence of understanding the whole process they are taking part. This, in turn, indicates the low level of farmers' empowerment through the program, a tendency which will be further discussed in Chapter 8.

Regarding the content of the training and the practices taught, several farmers expressed their difficulties with complying with some of them. The practices most often described as difficult to implement are the removal of mistletoes, pruning and the cutting of chupons. Especially women described their struggles with these physically demanding practices:

> "Everything helps. But something like pruning because I am a woman so I can't do it. Unless I find a man to help me. That is the only thing that worries me." (Society member in Wiawso type 3, July 2015)

> "With the mistletoes, they taught us that each time we must go through the farm to search for it and get them down but getting it down is not easy. If you climb up and you are not careful you can fall down and when you fall too you can die. And then also we need a kind of cutlass to remove it but there is none here so unless we use the same hoe that we have to use to pluck the cocoa and struggle to cut it, and if we don't cut it down the cocoa can go bad so you have to force it and cut it." (Society member in Juabeso type 2, June 2015)

Other interviewees highlight the danger of or their difficulties with the application of chemicals, particularly the spraying. In addition, chemicals are seen as expensive and their application is therefore challenging for them.

> "The most difficult is spraying because sometimes it enters your eyes and the heat that it comes with is too much. That's the most difficult but at the same time the most helpful practice." (Society member in Wiawso type 3, July 2015)

"Even mine I have not sprayed it because I don't have money, am very broke."
(Society member in Juabeso type 2, June 2015)

It seems that at the production level too, there is a distinct increase in workload under the certification scheme, an aspect which will gain some further attention in Chapter 8.

In addition to these difficulties some interviewees experience in their process of applying standard requirements, the possibility to get excluded from the society and be replaced by someone else is a pressure for those who struggle more in the process of application.

"When you receive the training as a member of the group but you don't follow it in your practice so maybe when you brought your cocoa it wasn't well fermented or something, you can be removed from the group and another person who is not in the group but is following the required procedures will replace you." (Society member in Bodi type 2, July 2015)

Finally, next to such group internal dynamics, the project brings a new dimension of inequality to the cocoa farmers in project-targeted communities and areas. Not only that participating farmers achieve a higher price for their beans due to the premium and also receive equipment which is not available to the non-participants, but also even the possibility to receive improved extension services and training and to have improved access to input delivery creates unprecedented differences between cocoa farmers in Ghana.

"When we get our clothes and shoes we wear and show to them and then they will be eager to join the group." (Society member in Juabeso type 2, June 2015)

"I wish I could register today so that when they are sharing the bonuses I can get some." (Non-participating farmer in Bodi type 2, July 2015)

"It is a problem because when they do not join the group they cannot learn anything. And so if they are all given the chance to come, it would help." (Society member in Juabeso type 2, June 2015)

In practice more than in farmers' own perception, this new gap classifies cocoa farmers in two different groups, that is, those who get the possibility to move forward and those who remain excluded from ongoing developmental processes in the area. The reason for this new fission among cocoa farmers in Ghana is the above described limitation in the number of participants. New discrepancies are present within a targeted community, between targeted and non-targeted communities of a same area, and finally between regions with high and low prevalence of certification projects. A dynamic which might lead to the creation of developmental islands, a dynamic which will be discussed further in Chapter 8.

The word cloud summarizes the main challenges derived from the interviews. The sizes of the words represent the frequency of their mentioning (Figure 7.2).

Figure 7.2 Word cloud of farmers' challenges. (Source: own elaboration)

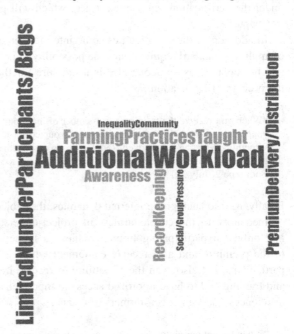

7.2.3   COCOBOD's Views on Private Sector-led Sustainability Interventions in the Sector

The views of the interviewed COCOBOD staff members regarding the increase of certification projects in Ghana's cocoa sector differed to a large extent among them. The following presented views only reflect individual attitudes towards and opinions on the increase of private sector-led activities and cannot be understood as a general or official position of the board. As presented in Chapter 5, the overall policy is to enhance public private partnerships through the CEPPP and to improve monitoring of private interventions through the PCU. The opinions of the interview partners range from positive attitudes which welcome the increased private engagement to skeptical views which consider this development rather problematic.

Positive attitudes

The effect of certification projects which is most often seen as a positive development in the sector is the expansion of training activities and outreach to farmers. Given COCOBOD's main objective to enhance farmers' competitiveness and provided the low ratio of extension agents to farmers in the country, several interviewees are in favor of any activities which contribute to the increase of extension services in the cocoa sector. The following quote represents such a view.

> "Yeah, it is really a positive step. Because COCOBOD can't do it all. So if you have them, the essential is to fill in that gap. The most important thing is the farmer to be more competitive. If the farmer is competitive, his productivity improves."(COCOBOD staff member, August 2015)

Some interview partners also differentiate between COCOBOD's operating range in the cocoa sector and the private sector activities which go beyond COCOBOD's activities by also tackling social and environmental aspects in communities.

> "So they bring them to the communities to train the farmers there on issues of gender to make sure that there is equitable distribution of resources between man and woman. We allow that. Because Ghana COCOBOD cannot do everything. Our core mandate is about cocoa. But all others we allow. If you have the funds to pay for why not, come! So it is not like a very straight jacket this is what we want you to do." (COCOBOD staff member, February 2015)

This statement represents two important positions. One, it becomes clear that the interviewee is convinced by the idea that private interventions have to be approved by COCOBOD, and second, that COCOBOD is open and flexible towards different approaches and interventions.

One interviewee observed a shift in private sector interventions which he identifies to increasingly put a stronger focus on the diffusion of GAPs. While generally welcoming this development, he also expresses concern about the business interest of the implementors.

> "Yeah, now I call it knowledge-based interventions. Yeah, because all what they seem to do now is to train farmers, to empower them to become more competitive. So as I

said, my thinking is, yes, they want the farmers to be more empowered, to become more competitive. However, they are looking to it with two eyes, because it is beneficial to them, they get higher revenue when it comes to that." (COCOBOD staff member, August 2015)

This view represents the perception of many COCOBOD staff members that TNCs would rather pursue their business cases when engaging with cocoa farmers. For some interviewees, this does not pose a problem and what counts for them is the positive effect on the sector. Others are more critical towards hidden business interests and particularly caution the capturing of the farmers through sustainability interventions.

Skeptical views
One interviewee, for instance, expressed a particular concern over the way how farmers are encouraged to apply GAPs. In his view, farmers are mainly motivated to participate through the premium and the other material benefits and do not really learn the intrinsic benefits of improving their production. Hence, he sees the sustainability of the intervention in question, as described in the following statement.

> "They will realize that their scheme of operation is not sustainable. Because farmers have rather been aligned because of certification. But if it is not there, what will happen? Will the farmers redraw again? That means we have not succeeded in training them properly. Because we have told if you do certification you will get extra premium. Now if the premium is not coming what happens? Or if the premium is reduced what happens? Will farmers continue with certification or not? And that is why we are a little careful with the way how things are happening. It is better to allow the people to make their own choices. But don't come and say our Europeans want good quality and that is the only way you can do it and we will give you a premium." (COCOBOD staff member, February 2015)

The point that the interview partner makes here is similar to the finding discussed in the previous section when discussing challenges with farmers' awareness on the objectives and process of the intervention. The awareness of the interview partner on this problem and the fact that he brings it up to discussion shows his good knowledge on the interventions and indicates an interest in farmers overall improvement.

While this interview partner has a general positive attitude towards the private engagement and rather questions the way how it is implemented, other interview partners look still more skeptical at the whole dynamic:

> "And to add on to this, you know, most of the donors are the chocolate manufacturers, the cocoa processors who also process the beans and their main reason for helping is to sustain the cocoa. So when you are asking how they term it, the objective is the sustainability of cocoa. So they just come in in order to give extension services to sustain cocoa, what can they do that there will be cocoa all the time for them. So sometimes they say, 'hey, let's give the people, the farmers, let's improve their livelihoods so that they can work hard to make sure that there is always cocoa. So the main idea is sustainability." (COCOBOD staff member, February 2015)

This statement represents an overall skeptical view on private sector-led sustainability interventions which are regarded to be mainly driven by profit and not by the goal to improve farmers' well-being.

One interview partner also identified sustainability interventions as an internal marketing tool with which farmers would be bound to the implementing LBC.

> "So the thing is they actually capture you as my farmer. When you are participating in a program you don't send your produce to any other place. So the larger the number of farmers that are captured the more cocoa we are getting, also for my mother company who is actually funding the project." (COCOBOD staff member, February 2015)

The capturing of the farmers is a recurrent point of view which many interview partners from COCOBOD challenge.

Another skeptical view that was expressed concerns the level of cooperation through CEPPP. As has been described in Section 7.1, COCOBOD provides its extension agents to the studied project and also to a number of other projects which are similar in its structure.

> "So I don't see it as a healthy partnership. Because, the partners at times come with their own interest. We should have a win-win situation but then they come with their own interest what are not always in line with our operations. For instance, a partner may come for certification and the certification entails a lot. And the community extension agents who are the implementers have to spend much time trying to satisfy...and they also have to satisfy the private partners which don't go in line with the CHED program or activities." (COCOBOD staff member, March 2015)

A degree of tension between the private and public sector perceived by the interviewee becomes apparent here. It is the core of CEPPP that COCOBOD provides logistics (hiring and training) and the private sector additional funding for extension services. For the participating TNCs, funding of a few extension agents is obviously not an issue and it is likely comfortable if COCOBOD hires and trains CEPPP extension agents for them. But it seems that for COCOBOD and the concerned extension agents who have to conform with multiple requirements, the practice of the partnership is not always easy.

Besides, some kind of competition between public and private sector seems to be experienced and the privileged position of private implementors due to their strong resource endowment is perceived.

> "But with a time, you go and you will get farmers out of that confusion: But the farmers are also interested in getting the financial support and we are not directly providing financial services they always tune themselves with them, the farmers always go the side of the privates. Because they come with the credits, they can provide that. We go there with services, organizing and training them but they need somebody to start up with, they want to get the input, and these people also go to the communities with the inputs and just educate them and they get them to their side. So you go there and you have nothing in the hand, but they come and tell 'we can provide you fertilizer, we can provide you financial support, we can give you cutlasses...The farmers need an initial capital to start something." (COCOBOD staff member, March 2015)

The two last quoted, skeptical views expressed by COCOBOD staff members from the district level feed their arguments from the practical side of and experiences with the implementation on the ground. In contrast, interviewees from the national level of COCOBOD stress CHED's central position in the extension field and focus on the need for improved coordination of private activities and their contents.

> "No, they don't really do that [seeking cooperation with COCOBOD, author's note]...that is actually where we come in, that collaboration is not there. For now they just go, because they know the LBCs have their farmers and go directly and then you know. So we are trying bring all of them on board and whatever they want to do, because when it comes to the technical support services, CHED should be the best person to give all that education. So whatever you want to do you come to us and then we see how best we can collaborate and support you." (COCOBOD staff member, February 2015)

As has been described in Chapter 5, COCOBOD's knowledge on private interventions in 2015 when PCU just started to work was still weak. Even if knowledge existed and information on private activities was increasingly gathered at the central parts of the board, at the local level of COCOBOD's various subsidiaries, the knowledge level regarding private sustainability interventions was very limited. The awareness on the studied project among CHED staff members in the district where the field work was conducted was very low. This is even the case having the the project's extension agent in the office. When being asked with which company the studied project project would cooperate, the interviewees did not know about it and it did not appear to be an interesting information for them. Also, there was no knowledge on other existing private sustainability programs at the district level. This, to some extent, shows the difficulties of COCOBOD to be able to efficiently oversee and monitor private interventions. A need that, if private interventions at the local level continue to increase like over the past ten years, will become key for COCOBOD if the aim is to not to lose public regulatory capacity.

When looking at the diverse perceptions of COCOBOD staff members regarding the rapid spread of private sector-led sustainability interventions at the local level, the ambivalence of the development becomes clear: COCOBOD staff members are aware that the board is limited in resources and CHED is unable to appropriately support the broad crowd of cocoa farmers, especially in the remote areas. The new engagement of the transnational private sector helps to achieve the goal of spreading extension services to more cocoa farmers. But there is a perception that business interests are hidden and also a diffuse picture, that the increased bounding of farmers to LBCs and their mother companies would not be good for farmers, is drawn. Furthermore, it seems that a feeling of competition between COCOBOD and TNCs is perceived by some interviewees. But because of the different positions in the GCCC, this perceived competition unlikely has to do with business competition. It rather seems to be linked to authority, such as responsibilities on the ground, relations with the farmers and sovereignty in decision making—all of which are directly and indirectly altered during the current dynamics.

7.3 Private Sustainability Interventions and Institutional Transformations in Ghana's Cocoa Sector

With the rapid increase in the implementation of transnational CSR strategies through sustainability certification schemes, Ghana's cocoa sector has undergone a number of changes in its institutional environment which affect the sector lastingly. Changes can already be observed looking at the individual level of one of such sustainability interventions as has been done by studying the project case. Yet, it is the sum of numerous similar and simultaneously implemented projects which leads to lasting sector changes. The main transformative dynamics that have been identified during the study are in the fields of sector organization, communication, and marketing relations.

7.3.1 New Forms of Sector Organization

The introduction of CSR and its sustainability interventions has brought a number of new responsibilities and activities in the sector which all together lead to new forms of organization of the local sector, most notably a tendential privatization of extension services, new networks of input delivery, new forms of farmers' organization and, the existence of an alternative sourcing system.

With the introduction of the IMS structure and the training activities linked to the UTZ standard requirements, a new approach of organizing extension activities has been established in the system. Although COCOBOD has been organizing farmers groups in order to train them with GAPs long before the start of sustainability certification in the sector, the way how it is done under the IMS is new and more efficient. The key tools are the 'train the trainers' approach and the working through a local resource person, that is the lead farmer. These new approaches and corresponding organizational structures transform the system of extension in a way that more farmers are targeted in a more resource efficient way. With this system, many farmers can be reached who, under the COCO-BOD extension system, received only seldom extension support, as the following statement underlines:

> "The time that I was facing the problems and I had nothing, the cocoa was bearing nothing, we had no one to teach us how to clear around the cocoa tree and how to prune and keep it, when we joined this group they taught us how to maintain the cocoa." (Society member in Juabeso type 2, June 2015)

With the implementation of the IMS, new positions with new tasks and responsibilities have been established in the system. With the increased involvement of LBCs in extension delivery, there is a strong trend towards the privatization of extension services. However, TNCs' involvement in extension services is likely not be regarded as a direct business opportunity but as a necessary interim step to improve training access for farmers.

Closely related to the IMS system are new relations between farmers and the private sector which change the access to chemical inputs for farmers. While under the studied project the attempts to establish links between farmers and input suppliers were still lose and in the interviewed communities these relations were not well established, yet the project lead firm in its interventions after the collapse of the studied project followed up to foster this concept apparently with more success. The input fairs which the company organized represented a completely new approach to input delivery in the system. Besides this new form of access to chemical products, the financing arrangement between cocoa farmers and a transnational processor is also new in Ghana's cocoa sector.

Moreover, there are new forms of farmers organization which haven't been in the system before the introduction of certification schemes. Farmers are obliged to meet twice a month to conduct the training activities. In theory they have to select positions in their groups and build up their organizational structure. But the extent to which these groups organize themselves and use the group for saving or other joint activities differs a lot among the communities. Nevertheless, it is a new dynamic which can trigger further developments of group organization in the sector. The following two statements indicate that some group activities above the training meetings are indeed taking place:

> "We buy our own chemicals, we use own chemicals to work but in difficult times of the group we tap into our coffers and we buy the chemicals and we all share to support ourselves." (Society member in Juabeso type 2, June 2015)

> "Yes we do discuss. When we meet we all bring our problems on the floor, we pave way for that and is a must for me and some of the members to go to farm to inspect." (Lead farmer in Juabeso type 2, June 2015)

The way how solid these new forms of organization among the farmers become and if there is a potential to grow them to the level of cooperatives has to be seen in future.

Since independency, the direct involvement of foreign buyers in cocoa purchases was not allowed. After the long period of purchasing monopoly of PBC and the liberalization of internal marketing, it was indigenous LBCs which were at the local market. But with certification, the trend that transnational processors or traders establish their own LBCs has increased. Still, LBCs with their processor mother companies such as Olam Ghana or Ecom Ghana continued to work through the local system of purchase through PCs. It is now the project lead firm that has done the first step to break this organizational structure and to get into a direct commercial relation with farmers. It is too early to tell whether this approach turns out to be successful and to trigger further developments towards increased direct buying in Ghana.

7.3.2 New Patterns of Sector Cooperation

All forms of certification implemented in the sector, be it in the LBC, FBO or NGO model, are established through new forms of cooperation among different sector stakeholders. In the case of the studied project, cooperation among partners became so strong that they entered a close alliance which for outsiders appeared as one joint actor, as the following statement by one COCOBOD staff member shows:

> "For a period of three years now they are expanding...and they come now out as an umbrella – a company or an entity." (COCOBOD staff member, May 2015)

This joint actor, materialized in the project steering committee where all strategic decisions were taken.

> "We do everything together. As a project manager, also [the project lead firm] has a project manager, he is also on the project, so we work along closely, so we decide together, and then we agree on every step we do. So it is a 100% agreement and we work hand in hand to implement the project." (Staff member of the project LBC, April 2015)

The partnership can be seen as a powerful entity which combines the strengths of the three partners. Therefore, this functional entity likely possessed several advantages in the field, as most visibly in the effective establishment of the complex IMS, something, that single actors as smaller LBCs are not able to do in the same way.

Without having gained experiences on the ground through the expertise of the partner NGO and the purchasing structure of the project LBC, it would have been more difficult for the transnational processor to implement its cocoa CSR project in Ghana on its own. The other most important form of new sector cooperation in the course of the studied project was the agreement with COCOBOD to provide extension agents to the project. This form of cooperation gave the project alliance the opportunity to benefit from the public knowledge and investment in training extension agents. This advantage is also used for the follow-up project. Extension agents of the studied project are important connectors between the private initiative and COCOBOD. The privately financed extension agents are located at the CHED district offices and have many interactions with CHED staff. These extension agents can therefore be seen as bridge builders between public and private sectors. When it comes to the implementation of community development activities, the project lead firm, similar to other cocoa processors and chocolate manufacturers, works through international NGOs. Here too, the partner is in the possession of skills and knowledges which are of benefit for the TNCs.

7.3.3   Shifting Marketing Relations

More generally, sustainability certification has brought the private sector closer to the farmers and companies are taking on new roles in the course of such projects. The marketing relationships between LBCs that implement certification projects and participating farmers have changed fundamentally as a result, as the following statements indicates:

> "The result of certification is that it has brought the trade very close to the farmers. I mean, the company has been working with farmers for a number of years, but they never really...their key focus...our key focus, we are just buyers, we buy beans, but now they have a sustainability manager, who is trying to see how can we get closer and build loyalty and support farmers directly and all of that." (NGO staff member, April 2015)

Furthermore, certification with all its associated advantages has become a central instrument of competition between LBCs. Since the benefits of participating in a certification program have become widely known among farmers, they prefer to sell their beans to LBCs that have such a project in place, even if themselves are not participants in the project. In this context, sustainability certification in recent years has most likely helped those LBCs with a certification project in place to gain market share. The following quotations illustrate this trend:

> "As I said, certification has become also a tool of competition. It is one of the things that, if you don't do, you lose, you lose market share. Because it has a lot of benefits

and then it attracts farmers. So that is why you will find that a lot of LBCs or a lot of groups are trying to do certification..." (IMS staff member, April 2015)

"Well, now, you know, the competition is very, very keen. Every company who is operating wants to participate on the UTZ. Everybody wants to do something which will enhance the farmers." (COCOBOD staff member, July 2015)

Moreover, several farmers started to feel more loyal to their LBC since the beginning of the training. In addition, bringing the produce to another LBC would imply to not to receive the premium. The following quotations highlight these trends:

"If you look at how they have helped us to get training in cocoa farming and you are able to get cocoa beans and sell it well you cannot possibly sell the cocoa beans to someone." (Society member in Juabeso type 2, June 2015)

"If you take it to a different place you will not get same premium." (Society member in Juabeso type 2, June 2015)

A PC's statement also shows this positive trend for the LBCs and his work:

"Yes more than at first. Before UTZ my work wasn't really doing well but since they came now even other people want to come and join it. Even those who are not part of the program are bringing their cocoa to me. So it's making the business go forward than before." (PC in Bodi type 2, July 2015)

Considering that the implementation of a certification project is very resource-intense and therefore mainly possible for the larger LBCs, it is very likely that there will be a trend of market concentration in Ghana's cocoa sector in the future. For smaller, indigenous LBCs, which are not able to run a certification project, it could become increasingly difficult to keep their regular customers with them, as they do not have comparable incentives.

In addition to this advantage, the traceability system also brings some advantages in another aspect for LBCs with a certification project in place. COCOBOD has designated warehouses for most LBCs in a certification arrangement in order to assure separated storing. In times when storage capacities are under heavy strain and trucks have to wait in line at the port, sometimes for months at a time, LBCs with certified beans can take their produce directly to the warehouses designated for them. The reduced waiting time saves them considerable costs. The estimation of a COCOBOD staff member illustrates this dynamic:

"Even if the warehouse is full, or there is a bit of challenge at the warehouse, the trucks of a certified LBC come, because it is traceable, they already have a reserved warehouse. And I think why most of the LBCs joined certification was from this thinking." (COCOBOD staff member, August 2015)

Ultimately, COCOBOD's external marketing relationships are also altering. COCOBOD has no influence on how many certified beans are produced and traded. Previously, COCOBOD was the only trading partner for international buyers. But now, when it comes to agreements regarding the quantities of certified beans, COCOBOD is only indirectly involved in the arrangement but has to guarantee a smooth flow at the ports.

"Yah, I think the certification is being run by the private sector. My point is, it is COCOBOD selling the beans but they don't control, they get all beans, whether certified or conventional, but they are not certain to say that Ghana can deliver 2000 or 100.000 tons of UTZ." (NGO staff member, April 2015)

The described shifting marketing relations are likely the most hidden within the various institutional sector transformations. There is strong reason to understand certification as a powerful tool for accelerated sector change. But, considering the presented dynamics, it has to be questioned whether sustainability can really be achieved with this way of implementing certification, if one considers strong tendencies of concentration and reducing public regulatory capacities as potential risks to the sustainability of the GCCC.

Finally, the implementation of certification projects in Ghana's cocoa sector has created a new product chain in the sector (cf. George Afrane et al. 2013). Next to the conventional chain, there is the certified chain now.[4] As described in Chapter 5, in the case of the conventional chain, TNCs do not have any link to cocoa farmers and only purchase the beans from COCOBOD, without knowing their origin. In the chain for certified beans, TNCs mainly from the processing segment have established a direct link, and by late have set-up their own local purchasing capacities, too. The legal requirements to obtain a license for local purchase by COCOBOD, and to buy the beans on the international market from COCOBOD are still officially met. Yet, in practice, in the certified chain, TNCs control the whole flow of the beans, have designated warehouses at ports and the selling to and re-buying from COCOBOD is only a bureaucratic act. Regarding

---

[4] There is also a chain for organic cocoa beans, mainly in the Eastern Region, but its importance in the sector is low. For a detailed study of the composition and functioning see Glin et al. 2015.

the delivery of extension services, COCOBOD used to be the main provider in the conventional chain. In turn, in the certified chain, extension in the frame of the standard training by IMS staff is the dominant form of extension service provision. The Figure 7.3 illustrates the restructuring of the cocoa chain in Ghana regarding marketing relations and extension services due to the implementation of certification.

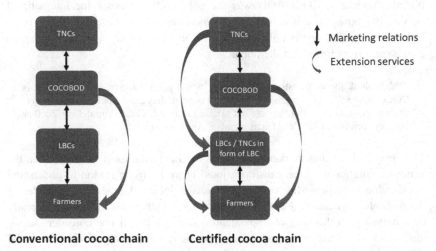

Conventional cocoa chain   Certified cocoa chain

Figure 7.3 Conventional and certified cocoa chains in Ghana. (Source: Own elaboration)

Transnational CSR as a Governance Tool and Local Development in the Global South

8

Sustainability certification has triggered diverse dynamics of change in the Ghanaian cocoa sector, which has been strongly targeted by TNCs' sustainability strategies over the past decade. While new forms of sector organization and patterns of cooperation can be observed at both the national and the local level in Ghana's cocoa sector, a major transformative process is most evident at the local level, and particularly in the high production areas in the country where most CSR interventions are located. In the Chapters 4 and 5, major sustainability challenges of the GCCC and more precisely at the local level of production in Ghana have been described. Thereby, CSR and sustainability interventions from TNCs have been presented as a tool to respond to a double pressure, namely the risk of a production shortfall due to increasing land scarcity and low levels of cocoa farmers' productivity in major West African producing countries, and social pressures which stem from a reducing legitimacy of the industry due to the high prevalence of poverty among cocoa farmers and environmental degradation linked to cocoa farming. This study has argued that the nature of this double pressure led TNCs from the GCCC to apply a governmental response which combines elements of industrial governance, hence control over local production, and soft tools of governance which aim to establish consent for the particular interest of the respective TNC. Many of these governmental activities are conducted in the frame of CSR/sustainability interventions. This chapter first brings the empirical results presented in the previous chapter in the context of the consent and control framework and thereby structures the empirical findings into governance tools of the transformative process and outcomes. Based on this structured information, the hypotheses will be discussed and research questions answered (Subchapter 8.1). Subsequently, these insights into local transformations linked to governmental interventions of TNCs in the form of CSR are brought into the

F. Ollendorf, *The Transformative Potential of Corporate Social Responsibility in the Global Cocoa-Chocolate Chain*, (Re-)konstruktionen – Internationale und Globale Studien, https://doi.org/10.1007/978-3-658-43668-1_8

superordinate context of the present study: Possible shortcomings and problems of the often articulated expectations towards TNCs to act as local development agents (cf. Barrientos 2011) are illustrated with the challenges and problems identified during the case study of one UTZ cocoa sustainability certification project in Ghana, and accentuated with some literature-based arguments. Thereby, the study turns back to its point of departure and seeks to contribute to the normative discussion of global sustainability governance and which role TNCs should play in it. Lastly, the discussion points to existing alternatives to TNC-driven local development in the Global South and highlights important avenues for further research.

8.1   Cocoa Sustainability Certification through the Lenses of the Consent and Control Framework

The "consent and control framework" has set out an ensemble of nine axes to assess governmental activities of TNCs through CSR. These axes comprise aspects of control and consent creation within a value chain and its institutional environment. They can be further divided into *process* dimensions, which help to describe the transformative implementation process of the governmental aspirations, and *outcome* dimensions, which inform the assessment of governance through consent and control. Accordingly, the axes actors, scales, topic of consent creation, and means of action are applied as process dimensions, and the axes geographical scope, input-output structure, industrial governance as well as institutional environment, and perceptions as outcome dimensions. Of course, axes are overlapping and often concern both, process and outcome of governmental activities, yet, the divide eases a systematic description. The following discussion is based on the empirical findings of the case study and is not directly transferrable to cocoa sectors in other countries or to other value chains. Particularly within the setting of large-scale farms, local dynamics linked to certification are completely different. Still, given the similarity of other sustainability certification projects in Ghana's cocoa sector, the discussion seeks to add some information to TNC-driven sustainability certification in a broader scope.

8.1.1   Dimensions of the Transformative Process

During the 2010-decade, sustainability certification was the major tool for TNCs from the GCCC to implement their CSR strategies. This has been likely the case because sustainability certification provides a clear structure and guidance on how to implement social and environmental standards, at the same time, through the multi-stakeholder development of sustainability standards and the third-party auditing, attaches a high level of legitimacy to procedure and product. The present study does not cover the processes of standards development, and hence, the discussion on how participative this process is and how strategic alliances act at this level is not part of it. Putting the focus of the study on the implementation level, it was shown how one major TNC from the GCCC established a strategic alliance with an international NGO and an important LBC in Ghana's cocoa sector because of its initial lack of local marketing structures and knowledge of local dynamics. Hence, the three different **actors** with similar interests merged into one strong sector player in the frame of the studied project. Each of the actors, the TNC, the international NGO, and the LBC, pooled their expertise and other resources to the project. Given the general lack of TNCs' experiences at the local level of cocoa production in Ghana and the absence of strong cooperatives through which they could reach to farmers, TNCs sought to ally with other organizations such as LBCs and NGOs who already have a link to farmers and who would establish the needed structure for the project. While certification projects are mainly run on a private basis and COCOBOD remained largely excluded from the steering and planning of implementation, in the frame of the CEPPP, COCOBOD opened up for new partnerships and provides its expertise to certification projects, mainly in the form of training of extension agents and offering working space at the district offices to them.

In the course of the case study project, participating actors were active at national and local levels of cocoa production in Ghana, hence working at multiple **scales**. The scale most strongly affected by the studied project and the sum of similar projects is the local level of cocoa production. Yet, at the national sector level, activities to shape the sector policy and to participate in decision-making at the various sector platforms, such as the Ghana Cocoa Platform or steering committees, are key activities of TNCs in the GCCC, too. The participation in sector dialogue platforms can also be regarded as an important means of spreading the meanings and values which foster the major interest of achieving the particular interest, that is the professionalization of cocoa farmers and the modernization of the cocoa sector which are supposed to lead to an increase in quality and quantity of cocoa production and an increased attractiveness of cocoa farming which motivates farmers to remain in the sector, hence *cocoa sustainability*. This major interest being framed in the notion of cocoa sustainability is the key **topic of consent creation**. Sustainability frames interests in a way with which all stakeholders somehow can agree (Leissle 2018, p. 177).[1] But what is precisely understood as cocoa sustainability, and how it can be achieved, what the sector priorities are and which policies are most fitting, are fields of a debate that from a TNC perspective needs to be steered towards the direction that fits most properly with its particular interest.[2] These are objectives which concern all companies in the GCCC and which can be regarded as their common interest which lies in a precompetitive field. Hence TNCs and their associations jointly advocate for them at the various sector platforms (interview with a high-level WCF representative in March 2015). In addition, the codes of conduct of UTZ and Rainforest Alliance have a distinct emphasis on training with GAPs which are regarded as major tools for the achievement of farmers' professionalization.

Seeking to push for the achievement of the professionalization and modernization within the sector, TNCs increasingly sought to intervene at the farmers level. As described in Chapter 5, after the introduction of the SAPs from the 1980 s onwards and the linked drastic reduction in public expenditures in cocoa sector support, a huge void in service provision to cocoa farmers arose. One major challenge for TNCs from the GCCC in Ghana used to be the missing link with cocoa farmers since they did not have marketing relations with them in place. Hence, CSR in the form of sustainability certification provided the companies with new tools to overcome this gap. These tools represent what the consent and control framework captures with governmental **means of action** and combine aspects of production control and consent creation. The implementation of sustainability certification projects such as the studied project comprises these two dimensions. With the establishment of the IMS structure, a sophisticated enabling management structure (initially called "internal control system") is set up. Typical industrial governance tools are applied to control participating cocoa farmers, to which without the project, such tools would not have reached. First and foremost, the contracts, spelling out farmers duties and benefits, are a concrete measure of industrial governance. Even if farmers are in many cases not aware of the contract details or even the existence of it, there is a clear understanding among them that certain rules, which the IMS staff members have called the "three basic laws", that is attending the meetings, following the code of conduct, and selling the

---

[1] The functioning of such policy terms as empty significant has been described by Cornwall and Brock 2005.

[2] Cheyns and Riisgaard have described a process of the exclusion of more civic claims from the concept of sustainability in the context of multi-stakeholder sustainability platforms: While addressing health and working safety issues or the improvement of yields, claims for more equality in access to resources or value distribution are neglected in the discourse, Cheyns and Riisgaard 2014, similarly argue Neilson and Pritchard 2007.

beans to the LBC of the studied project, need to be respected if they want to be part of the project society and receive the premium. With this requirement for alignment with the project rules, the relations between the farmers and both, the LBC and the TNC, alter and farmers get somehow bound to the two project companies. There seems to be no immediate legal sanctioning for farmers who would decide to sell their produce elsewhere. But the fact that the farmer societies do not hold the certificate on their own but the project LBC prevents farmers from the opportunity to sell their certified beans as such to other LBCs.

Besides, the UTZ Code of Conduct, combines detailed requirements regarding production, post-harvest treatment, and farm management on the one hand with social and environmental practices on the other hand. However, most of the interviewed farmers perceive a greater importance of process standards compared to the social and environmental ones, indicating a stronger focus on the first during training sessions. Through the IMS and the 'train the trainers' approach a highly efficient way of diffusing extension messages to farmers is established. Similarly, the new systems of access to chemical inputs support the modernization of cocoa farming and the increase in production rates and can be seen as industrial governance means of action, too.

In order to increase control over application of and compliance with requirements, sophisticated monitoring systems, most notably record keeping and auditing, are established. In the studied project much documentation and reporting were required by UTZ and new means of digitization of farmers' and farm attributes were being tested. Moreover, with their participation in the certification scheme, farmers agree to document all their farming details and deliver the information to the project. A close supervision of the adaptation of practices is done by lead farmers, IMS officers, and finally by external auditors. Next to these new forms of monitoring of farmers' activities, control over the adaption of standard requirements is also achieved by two sanction mechanisms: the exclusion from the society in case of repeated non-compliance and therewith the loss of the premium and the other benefits. These new forms of monitoring and sanctioning in the sector represent new mechanisms of industrial governance which come along with the accumulation of data and other knowledge by the TNCs, which represent another governmental advantage for them.

In combination with these industrial tools of governance through increased control (contracts, standards, and monitoring and sanctioning mechanisms), important means of action to foster consent on the need for cocoa sustainability, understood as farmers professionalization and sector modernization, are applied. As unrolled in the theoretical chapters, important governmental elements to foster consent comprise compensation and material concessions, dialogue, and creating enabling environments. Such measures can be identified when analyzing the project lead firm's activities linked to CSR in Ghana's cocoa sector. These times, the risk of a real uprising of cocoa farmers because of low prices and their poverty as it was lastly the case in the 1940 s does not seem very likely. More likely seems their reorientation to other cash crops such as rubber or palm tree production or their complete exit from farming if prices continue to be too low to enable a decent living income. While empirical findings rather indicate that farmers' actual interest in cocoa as a future livelihood strategy depend from their educational and socio-economic backgrounds (cf. Anyidoho et al. 2012), TNCs being dependent from the willingness of cocoa farmers to continue to produce, fear their reorientation towards other livelihood strategies. This was underlined by the quotes of staff members of the project lead firm in Chapter 7. In Chapter 7, the delivery of the premium and equipment was presented as an important incentive to make farmers join the project. But next to this function, the delivery of premium and minor production assets can also be regarded as important means of material concessions seeking to prevent farmers' reorientation.

It seems that the delivery of premium and equipment, even if only marginal material improvements for farmers occur, also serves the purpose of compromise in order to prevent farmers exit of the sector. Another important function of material concession is fulfilled by the program of community support, which plays a more strategic role in the recent cocoa sustainability strategy of the TNC as it used to be the case during the period of the case study project. As has been clearly stated by a staff member of the project lead firm (Chapter 7), the objective is to prevent farmers from migrating to urban areas and thereby drop out of cocoa production. Hence, measures of community support explicitly seek to respond to this challenge by improving farmers' living conditions as much as needed to prevent a broader reconfiguration of the system.

Besides, in order to foster consent for the hegemonic project, TNCs seek to co-engage in the creation of an institutional environment that facilitates the modernization of the sector. Many of the studied TNC's activities linked to its CSR strategy can be assessed from that angle, too. For instance, all described means of actions which concern innovations in the marketing structure, the establishments of new input delivery schemes or the implementation of the 'train the trainers' and approach, as well as the installation of the IMS, are proactive interventions which seek to shape the institutional environment of Ghana's local cocoa sector. In a similar way, the attempts to establish new mechanisms to improve input access, by the creation of a direct link between transnational input suppliers and farmers, are concrete measures aiming to overcome local institutional shortcomings and therefore have to be understood as governmental undertakings by the transnational processor.

Finally, another important means of action for consent creation is the implementation of new forms of formal or informal dialogue or the participation in already existing ones. Dialogue can be regarded as one of the most direct forms of spreading meanings and values and builds the basis for cooperation with other stakeholders. In the course of the studied project, dialogue is practiced at various scales, as already indicated above. There is a daily exchange between extension agents and CHED staff members, there are discussions on sector policy meetings, there are many formal and informal meetings along the whole IMS hierarchy where dialogue takes place. It is very likely that the understanding of cocoa sustainability as the modernization of the sector, which is inherent to the approach of the studied project, is communicated during the various occasions of sector dialogue.

8.1.2 Outcomes of Private Governmental Aspirations through CSR

Having assessed the CSR activities of one major TNC from the GCCC through the lenses of the consent and control framework, a broader set of governmental means of action came to light than with the description of the main fields of action during the implementation process (Subchapter 7.1). The consent and control framework equally allows to dig deeper in the outcome dimensions. The axes input-output structure, geographical scope, industrial governance, institutional environment, and perceptions help to further conceptualize governance achievements for the project lead firm and other TNCs in the GCCC linked to sustainability certification. Thereby, the study is limited by the scope of the empirical insights gathered during field work. For some of the axes, more empirical data would be required in order to provide a robust answer. Before entering the discussion, it seems important to highlight again, that the following discussion is based on qualitative research findings, that is interviewees' *perceptions* of effects regarding production and sector dynamics in 2015 and 2017. Hence, the discussion cannot be taken as a robust basis for generalization but rather shows tendencies which became apparent during analysis.

Regarding the **input-output structure** of the CSR intervention, several patterns are emerging. First of all, even if still weak and experimental, there is a better access to input factors for participating farmers. The difficulty in obtaining good quality inputs and the many fake products on the local market were among the challenges most often mentioned by the farmers. In 2015, the interview partners complained that the project LBC and the lead firm had made promises to help farmers get better access to chemicals but they failed to honor it. In 2017, the project lead firm had taken up a new approach to improve input access and begun to arrange yearly input fairs which apparently were successful. In addition, the delivery of Personal Protection Equipment and spraying machines, financed by the project sustainability premium, is another new form of input delivery introduced by the certification project.

Looking at the output side, perceptions of interviewees from all stakeholder groups point to the same direction that both, quality and quantity considerably improved with the participation in the project, which would present a major governmental achievement for the TNC. Yet, literature findings are inconclusive: While some researchers (Ingram et al. 2018; Iddrisu et al. 2020; Fenger et al. 2017; Brako et al. 2020; Waarts et al. 2015) identify productivity increases for certified farmers, Ruf et al. (2015a) uncovered "structural contradictions" such as the "scarcity of labour and underestimation of risks" (Ruf et al. 2015a) within certification which would limit positive production effects. A look at the production rates for the Western region, which has been most strongly targeted by certification programs over the past ten years, supports the argument of only marginal effects of certification programs on farmers' production rates: No clear trend of productivity increases in the Western region can be spotted for the past ten years. Rather, production varies from around 500,000 tons in high to a bit less than 400,000 tons in low production years (Ghana Cocoa Board n.d.), as the following Table 8.1 shows. Hence, the effect of certification on cocoa production over the past decade remains inconclusive. Either, not enough farmers are targeted to see an effect on the regional scale, or, since Ruf et al.'s findings in 2015, productivity gains have not increased further.

Table 8.1 Evolution of cocoa production rates in Western region, Ghana, in metric tons

Year	2011/12	2012/13	2013/14	2014/15	2015/16	2016/17	2017/18	2018/19
Production	525,237	458,107	482,691	380,469	415,302	496,224	453,992	364,383

Source: own elaboration based on Ghana Cocoa Board 2020.

The **geographical scale** of the distribution of CSR projects almost exactly reflects the position of the cocoa producing regions in the national cocoa sector as can be seen in Table 8.2. The Western region, being the most productive cocoa region in the country, is targeted by the highest number of sustainability interventions, as per PCU report in 2014, followed by the Ashanti region, second most productive and equally second most CSR-targeted region.

Table 8.2   Number of CSR projects against cocoa production per region in 2014

Rank	Region	Production in 2013/14 (metric tons)	Number of CSR projects in 2014
1	Western	482,691	26
2	Ashanti	156,871	19
3	Brong Ahafo	87,050	12
4	Central	85,435	11
5	Eastern	80,692	9

Sources: own elaboration based on Ghana Cocoa Board n.d. and Project Coordination Unit 2014.

The pattern shows that investment obviously goes first where highest returns can be expected and production increases are still even possible in the high productive areas. These figures tell the reader that cocoa farmers in high production areas seem to have better chances to receive training and other services and benefits in the context of a certification program than farmers in the other regions. However, there are no clear figures on the number of cocoa farmers per region and, hence, a direct conclusion of whether this trend leads to spatial fragmentation of development cannot be drawn. Possibly, in the other regions, the number of cocoa farmers is smaller than in the high productive areas and hence, the ratio CSR-targeted farmer to non-targeted farmer must not necessarily be too different between the regions.

With the implementation of the studied project and subsequently of the TNC's cocoa sustainability program, new forms of marketing relations are established, introducing new links between producers, suppliers and buyers, and implying new forms of **industrial governance** relations. In the course of the studied project, the TNC entered new marketing relations with farmers, an LBC, and COCO-BOD. The main achievement is the direct link with Ghanaian cocoa farmers which was inexistent before. This has extended the TNC's sphere of influence in Ghana's cocoa sector down to the production level. As highlighted above in this chapter when describing the governmental tools, contracts and potential sanctioning mechanisms of exclusion bound targeted Ghanaian farmers in an unprecedented way to a lead firm of the GCCC. While farmers before their participation would decide to either change their LBC or stick to it because of loyalty, for a first time they are officially bound (although not legally pursued in case of non-compliance) to sell their beans to the project LBC. The relation between the TNC and the project LBC was functional and both firms benefitted from the joint certification scheme. Yet, the risk of damaging marketing relations with farmers' due to their disappointment with the project was solely borne by the LBC. After the collapse of the latter, the TNC opted to go on its own, indicating that a backwards integration in this case is more efficient for the company than a joint venture with another LBC. In addition, the marketing relation with COCOBOD has changed, too. As has been presented in Section 7.3., COCOBOD plays a rather secondary role in the marketing of certified beans and only facilitates the sound flow of the certified produce. This trend has potential to weaken COCOBOD's position in the GCCC since the lead firms' targets are to purchase 100% certified cocoa and has to be further investigated carefully.

Ultimately, all these changes linked to the introduction of sustainability certification together have transformed the institutional environment in Ghana's cocoa sector in a way that the TNC has expanded its field of influence and control. The overriding **research question 1** scrutinizes if CSR serves TNCs as a tool to extend their power position in a given value chain. This overriding question will be answered based on the answers to its two sub-questions. **Research question 2** asked about the tools and mechanisms of how a CSR strategy, in our case in the form of a sustainability certification program, is implemented and how these activities lead to the transformation of the local institutional setting in Ghana's cocoa sector. The empirical study of one UTZ certification project unveiled the introduction of a highly sophisticated, hierarchical management system, which is a requirement by the UTZ Code of Conduct, and with which many new responsibilities and positions in the sector were created. Major transformative effects have been presented here, such as the tendential privatization of extension services, new networks for input delivery, new forms of farmers' organization, new alliances between stakeholders and new forms of cooperation, as well as changing marketing relations, which put a new pressure on smaller local LBCs. In this context, certification has been described as a competitive tool among LBCs which has contributed to the entry of TNCs in the form of LBCs in Ghana's local cocoa marketing system. For the case of the lead firm of the studied project, the TNC has increased its efforts at the local level over the past few years. It is experimenting with different approaches to change the local system, such as different measures of farmers training, distribution of inputs or the sourcing system, which is a specific of Ghana's cocoa sector since independency. In sum, through the implementation of its CSR strategy, a multitude of local transformative processes which aim at supporting farmers' professionalization and sector modernization, hence fostering the TNC's particular interest, were triggered. These processes are still ongoing and place the company in a position of an important local sector player. This, however, is a development which affirms the hypothesis 2, that CSR interventions transform the institutional setting in a way that enables TNCs to gain more control over the fragmented production base in Ghana.

Research questions 3 asked how CSR affects cocoa farmers' perceptions towards their livelihoods and cocoa production, hence, if cocoa farming in the course of their participation in the certification project became more attractive for them. Moreover, other stakeholders' perceptions towards the role of TNCs at the local level were also of interest. The presented empirical findings indicate that effects are minor here. Indeed, farmers' interest in participating in a certification project drastically rose over the past decade. However, their major incentives are by large the delivery of premium and material equipment, followed by the vision of easier access to inputs, and participation in the training[3] and rather not a perceived improvement in their livelihoods. Many interviewed farmers attach a sense of modernity to certified cocoa and the "new farming practices". These seem to be good conditions to achieve the main governmental interest of professionalizing farmers. Still, to date, the additional labor efforts and time investments for certification of interviewed farmers have not translated into a significant improvement of their living conditions. Hence, most of the interviewees do not foresee the future of their children in the sector. Only the most vulnerable farmers, which stated to have little or no educational background, see cocoa as their major livelihood strategy. These findings are similar to research findings in the cocoa sector before the sustainability certification boom in 2010. It therefore seems most conclusive that, given the little economic improvements that occur to participants, the intervention is not able to capture the farmers in a way as desired by the TNC and that the risk of their reorientation still persists. This explains the TNC's further elaboration of its cocoa sustainability strategy and intensification of measures, as for instance to now put a stronger emphasis on community support activities, too.

Looking at COCOBOD staff perceptions regarding certification in the sector, the picture is diverse, too. On the one hand, the new forms of sector cooperation and dialogue have brought public and private sector closer to each other and policy deliberations are in a multi-stakeholder arrangement now. The TNC's and the international NGO's staff members emphasized their contentment with the fact that COCOBOD is increasingly open for cooperation. Still, while some interviewees from COCOBOD were satisfied with the improved situation of service delivery for farmers, others were more skeptical towards the proactive role

---

[3] Deppeler et al. 2014 obtained similar results in their interviews with certified farmers in Ghana.

of the TNCs at the local level, and, assuming the business-drive of these inter-ventions, cautioned that farmers would get captured and become dependent from TNCs. Hence, the hypothesis 3, that CSR helps TNCs to create consent for their particular interest and to achieve sector modernization while leaving critical dis-tributive asymmetries in the GCCC untouched, can only partially be affirmed. Nevertheless, given all this, the **research question 1**, whether CSR serves as a tool to enhance power position of TNCs, is still answered positively. The case study of one UTZ cocoa certification project and insights in subsequent dynamics under the project lead firm's sustainability strategy have shown how one of the lead processors in the GCCC implemented new measures of control over cocoa farmers production activities. The now drastically increased interest of farmers in certification projects and a tendential sympathy of COCOBOD staff with sustain-ability interventions show that consent for TNCs' local activities is also spreading. However, this process of gaining control over the local production base and of holding enough consent for this project is still ongoing and will remain dynamic. TNCs are learning and developing ever more specific measures to gain more and more ground in production areas which are difficult to access. CSR has become a complex system of sustainable supply chain management which does not hide the business case model anymore.

8.2 TNC-driven Sustainability Certification and Local Development Opportunities

8.2.1 Regulatory Capacities at Risk

Since the rapid spread of sustainability interventions at the local level of pro-duction, COCOBOD faces difficulties to efficiently oversee and regulate these dynamics, since till now, besides the training of extension agents, it is not involved in a relevant way in the conduct and monitoring of certification schemes. Given a continuous increase of private activities on the ground, COCOBOD's position in the sector will likely further weaken. This risk of a lasting decrease of public regulatory capacities in CSR-targeted countries in the Global South due to the establishment of TNCs' CSR strategies has been already stressed by Tallon-tire (2007, p. 779). As described in Chapter 7, the chain for certified beans largely by-passes COCOBOD's extension structure and COCOBOD is obliged to accom-modate the traceability requirements in its conventional marketing structure. The need to integrate these new demands in its own structure and the above described changing quality of marketing relations with TNCs elucidate that COCOBOD is not on top of the decision-taking process anymore and to some extent lost its sovereignty in the sector. That does not mean that the marketing board does not have the official role as principle sector regulator anymore. But the dependency from good relations with international buyers and their demand for certified beans urges it to be open towards certification and provide the needed infrastructures.

Initially most of the CSR interventions were established in the field without any involvement of the public sector and often without COCOBOD being aware of it. The creation of Project Control Unit (PCU) can therefore be interpreted as a counter-movement step seeking to regain control or at least monitoring capacity over local activities. While these efforts are challenging and to some extent depending from TNCs' goodwill to share information, there is a new trend, too: The private sector increasingly wishes COCOBOD to assume a regulative and facilitating role in the sector but to rather refrain from its involvement in service delivery (see quotes in Section 7.1.). The quotes presented in Chapter 7 show this changing perception of TNC and WCF representatives towards the public sector's role. Now the common understanding seems to be that it is more efficient to use the resources of the public sector rather than by-passing it completely, as initially attempted. Still, the private sector has a clear idea of what role COCOBOD is supposed to play in the modernized sector and it has to be seen in future to which re-configurations in the sector these expectations lead. In any case, an efficient oversight and regulation of private activities would be a strong challenge for the board. Even though having PCU in place, participation in CEPPP and reporting on the alone-standing activities are voluntary. Hence, COCOBOD currently has no appropriate information basis to assess and regulate private sustainability programs in the sector. And the trend of this information asymmetry is just increasing: In the course of their interventions, TNCs and certification organizations dramatically extend their information bases on local production conditions. While the possessing of such supplier information constitutes a real advantage against competitors in the value chain (cf. Ponte 2019, p. 18), it also improves the role of TNCs vis-à-vis the public sector which, partly due to its limited extension resources, by the time of writing of the present study, struggles with appropriate data collection. Finally, the possessing of production information provides the basis for sound sector planning in which COCOBOD might fall behind in the process of public-private sector policy negotiations if it is not expanding its own data basis. Furthermore, the digitization of farmers' production information, for instance through the introduction of farmers' ID cards as done by several cocoa sustainability projects, hence, the development of block chain capacities in the sector, will likely further contribute to increasing power asymmetries (cf.

Mooney 2018) in the GCCC since the ownership of production information is a key asset in digital supply chain management.

8.2.2 Upgrading Opportunities Reduced

Power relations are also changing at the local level of the GCCC. In Ghana, smaller LBCs with less resources to implement the logistics for the demanding IMS structure are facing new forms of competition—either with the already existing strongest LBCs or newly entering TNCs in the internal marketing segment. This trend is similar to the "sustainability-driven supplier squeeze" (Ponte 2019, p. 18) that Ponte observes in dynamics of further consolidation promoted by lead firms in their supply base. In addition, certification as a new tool for competition between LBCs as well as farmers' increased interest in participating in a certification project because of their wish to receive the associated benefits also raises entry barriers for new local LBCs. With the increase of TNCs' share in local marketing returns, Ghana's potential to upgrade its economic benefits from the cocoa sector is likely reduced since these gains do not remain in-country. Furthermore, in the case of the studied sustainability project, a whole occupational group, that is the purchasing clerks, is squeezed out of the chain. At the same time, cocoa producers' upgrading in the chain due to their participation in a certification project is unlikely, too. While it is difficult to collect robust data on certified farmers' time expenditure on standard requirements, several studies point to the increasing workload for them coming with production techniques such as pruning, the removal of mistletoes and chupons, and post-harvest handling (Lemeilleur et al. 2015; Deppeler et al. 2014). This makes it also difficult to determine the costs of compliance, which are also dependent from various other factors such as farmers' material endowments or technical levels and size of farms (Lemeilleur et al. 2015). What seems to be the case is that farmers' additional efforts during post-harvest handling translate into improved quality of the beans, but due to the fixed price for cocoa beans in Ghana, this does not lead to a better farm gate price for them. In contrast, it is rather doubtful that the additional workload is in all cases covered by the premium. As Neilson and Pritchard (2007) criticize in their study on coffee certification in India, the lack of public regulatory capacity makes it even impossible to regulate who finally bears the additional costs and risks associated to the compliance with certification standards.

8.2.3 No Fairness in a Certified Cocoa Bean

It becomes clear that, during thes implementation of the sustainability standard UTZ by the lead firm of the studied project, the focus is mainly on quality and quantity aspects of production, out of which the majority are irrelevant for an environmental sustainability, and marginal for the improvement of social sustainability challenges (cf. Lemeilleur et al. 2015). Moreover, even if ILO core labor standards are included and farmers officially required to apply them to their farm workers, the voluntary approach of the scheme prevents the application of workers' rights to lead farmers and PCs. These important key actors in the IMS are regarded as voluntary resource people—rather than IMS staff, and therefore do not receive a remuneration which would be appropriate to their efforts and time expenditures. On the contrary, they are expected to contribute their time and efforts voluntarily to the scheme without making any benefits from it. Similarly, farmers' coaches from the follow up sustainability project receive very low allowances for their activities. Interviewed IMS staff members did not provide information on their allowances. But the fact that, in order to motivate the IMS staff of the studied project to continue with the LBC of the project lead firm, the project lead firm offered the double amount of salary than its competitor to the IMS staff, indicates that in many cases, IMS staff's incomes are low, particularly for those at the bottom level of the hierarchy. Deppler et al. (2014) stress the absence of transparency regarding the distribution of the premium at several levels within the certified chain. During the field work of the present study, another factor of lacking transparency that became apparent was that farmers were not aware that the material benefits they received were actually paid by the premium, hence by their additional work efforts. Rather, the LBC used the distribution of PPE (boots, overalls) in a way that it appeared as a gentle gift to the participants.

Finally, looking at the living conditions of the farmers who participated in the studied certification project for already four years, and listening to their major concerns and needs, it seems inappropriate and misguiding to meticulously calculate the supposed increase in their revenues as done in most of the few existing studies on sustainability certification in Ghana's cocoa sector. These farmers visibly still live in severe poverty and even if interviewed farmers saw their production increase with their participation in in the studied project, none of them reported that this production increase in combination with the premium for the certified beans would have enabled them to escape from poverty. After four years of participation, building a concrete house or assuring school fees were still challenges for them. It is a fact that the premium is far too low to assure sustainable living incomes (Hütz-Adams and Greiner 2020).

While still holding to it, the industry is aware of this major shortcoming of certification. Already in 2015, a high level WCF representative in an interview did not hide the failure of certification to reduce poverty:

"Well...let me put it this way – certification is not giving what it was originally designed to do. The theory is not holding up in the reality and we just state it like that. It doesn't mean that the companies are not buying certified product, it doesn't mean that the companies haven't made public commitment to go 100% certified... Many of them have. But, even if we have 100% certified...Ghana will still be full of poor farmers. So, the issue is, the way at which we look at certification, it is a tool in a toolbox but it is not enough to change the sector." (High-level WCF staff, March 2015)

Similar statements were made by the interviewed staff members of the partner NGO:

"In spite of certification farmers are still poor. So, are we certifying poverty? People are raising questions." (Staff member of partner NGO, April 2015)

Hence, there seems to be a general awareness among industry members that certification is failing to reduce, let alone erase poverty of cocoa farmers. Certification being seen as only one tool in a toolbox to achieve sector modernization, most TNCs, under the umbrella of the WCF, join efforts in the precompetitive fields of farmer livelihoods, community empowerment, human rights, and environment (cf. World Cocoa Foundation n.d.a), and further elaborate their CSR strategies as has been shown in Section 7.3. But when following the goal of sector modernization, one key aspect is completely ignored by the industry, whose main interest lies in the lasting and sufficient supply of beans: What will happen if modernization efforts finally become efficient and production significantly increases? Already in 2016, a slight oversupply of only some percentages more than the global demand led to a sharp decline in cocoa world market prices (Hütz-Adams and Greiner 2020) causing an income decline of 30%-40% for cocoa farmers (Fountain and Hütz-Adams 2018), though not in Ghana, where cocoa farmers are protected by the fixed price set by COCOBOD. Fountain and Hütz-Adams (2018) describe a "structural oversupply that could last for years to come (Fountain and Hütz-Adams 2018, p. 8). They see two main reasons for this development: (1) a large number of new farms which were established in protected forest areas mainly in Côte d'Ivoire, and (2) the industry's growing efforts for productivity increase in the form of sustainability programs. Yet, this aspect is not covered by either industry-wide sustainability strategies, TNCs' in-door sustainability programs, or sustainability certification schemes. A continuous significant increase in cocoa production due to sustainability interventions risks to lead to such low world market prices and hence farm gate prices that would further impoverish cocoa farmers until a level where most vulnerable ones would drop out of the sector. Given the likely increase in TNCs' efforts to boost production at farm level, this scenario needs to be addressed urgently.

8.2.4   Shortcomings in Underlying Developmental Assumptions

This need brings the present discussion to the underlying developmental assumptions being inherent to the private sustainability schemes. The "productivist rationality behind a sustainable certification process" (Lemeilleur et al. 2015) bases on the major assumption that the increase in quality and quantity would automatically translate into a higher income and therewith better social and environmental conditions for cocoa farmers. In addition, the code of conduct comprises important social and environmental aspects of cocoa production conditions, such as the ban of child labor and the protection of local water bodies which often serve as drinking water sources. While the real impact on economic, social, and environmental capitals remains unclear to date, it has been argued that the UTZ Code of Conduct fails to respond especially to those components of development, which are perceived as most important by targeted cocoa farmers for the improvement of their livelihoods (Dengerink 2013). In Dengerink's study, these were aspects of human and physical capitals, such as health or working conditions, and quality housing or sanitation respectively, which are similar findings as the ones presented in Chapter 7. As discussed above, the growth in production does not allow to increase the income to such an extent that fair living incomes, which would allow to improve these developmental dimensions, can be achieved. This would require a much greater shift in the distributive pattern of the GCCC. But it is intrinsic to CSR promoted by TNCs to not tackle underlying structural reasons for poverty such as unfair distributive patterns, which reproduce asymmetries in the global value chain (Tallontire 2007). Because of this very nature, CSR indeed can serve as a strong tool to provide short-term benefits for targeted farmers. Skill development and capacity building through improved extension schemes are definitely a positive outcome and might also trigger some more lasting effects. But CSR fails to address the broader long-term perspective of local development by disregarding systemic challenges. This becomes particularly evident looking at CSR's complete ignorance of the need for alternative livelihood strategies for those farmers not able to modernize and compete, and hence doomed to drop out of the sector. Besides, given the fact that CSR is an exclusive approach and only targets some particular groups, it should not be mistaken for a local development approach. While being less the case in Ghana's cocoa sector, several studies have shown how especially better organized groups of farmers, and particularly not the most vulnerable or marginalized ones, are most benefitting from certification projects (Neilson 2008; Ruf et al. 2019b). In addition, being "captured" in a certification scheme of which TNCs retain ownership (cf. Neilson 2008), farmers receive full service packages, comprising access to credits, new input delivery schemes, extension services, local sourcing—all facilitated by the same LBC or TNC which runs the project. In a context of internal marketing consolidation and weak local input markets, hence, the absence of choice, the risk that farmers become dependent from strong LBCs, which increasingly are subsidiaries of the transnational processors, is very high.

It has been argued that extension services through private or public-private sustainability programs are more farmer-driven than the "top-down" public extension in Ghana's cocoa sector (Laven 2010). While it is definitively the fact that the "train the trainers"-approach and conduct of training sessions by local lead farmers is very efficient in remote areas with weak infrastructure, the level of self-direction and process participation can be questioned with the findings presented in Chapter 7. While COCOBOD's extension structure at least rests on an institutional system developed by a legitimate government, the IMS is equally organized in a strong top-down structure but is steered exclusively by the (transnational) private sector. Its accountability mechanisms are only directed downstream whereas no accountability mechanisms for downstream players towards targeted farmers are foreseen. Moreover, in the IMS structure of the studied project, farmers are not represented in the higher levels, do not have any stake in decision-making processes, and are only the bottom link in the chain that is supposed to receive the program and fulfil requirements. The IMS approach of the studied project visibly fails to meet critical criteria of even a basic definition of participation which foresees "the participation and involvement of (ordinary) members of a group, an organization, etc. in both, setting and realization of its goals" (Fuchs-Heinritz 2013, p. 489, author's translation). Given the lacking inclusion in the IMS, the main drive of farmers coming together for training sessions is the requirement for it if they want to obtain the premium. The interviewed farmers did not state any sense of self-organization going beyond this practical need to gather on a bi-weekly basis. Thus, the trend that TNCs increasingly engage in the formation of farmers organizations or even cooperatives, which they see as best form of farmers organization to distribute extension messages and implement traceability schemes, should not be confused with a real farmers empowerment, where emancipated farmers come together to assess their needs, deliberate on best strategies, take informed decisions and represent themselves politically. But indeed, looking through the lenses of the consent and control framework, the implementation of such "pseudo-participative" structures, can serve to prevent the formation of real participatory structures. In fact, such structures would be an important component of the counter-hegemonic process of a grassroots resistance which follows the goal of transforming the poverty-producing and marginalizing structures by the people affected by them (Leal 2007, p. 539f.). Furthermore, with insufficient participation of the target groups, the auditing of the fulfillment of implementation of the standards by auditing firms can be seen as a highly paternalistic process (cf. Levy and Newell 2002). In the presented case study, auditors enter the cocoa farm and inspect it, judge the degree of adaptation, and decide on whether the farmer will pass the audit and get certified or not. A strongly asymmetric process in terms of power positions can be identified in this practice. Notwithstanding, as some authors argue, the implementation of standards is often discussed as a neutral technical process. By doing so, a more power sensitive look at it is actually hampered. This in turn inhibits the articulation of critique on existing asymmetries in developmental processes and ultimately undermines a real struggle over structural dimensions of business-poverty relationship and structural change (Levy and Newell 2002; Blowfield and Frynas 2005; Blowfield and Dolan 2008).

Major deficits and challenging outcomes of private sustainability certification have been presented here. Among the most critical ones are the undermining of local and national regulatory capacities, an increased tendency of consolidation in the internal marketing sector to the detriment of local Ghanaian LBCs, the continuous poverty among cocoa farmers, the risk of overproduction leading to a world market price decline, and a merely functional approach to farmers' participation which could even hinder real emancipatory developments at the local level of Ghana's cocoa production. Given all these arguments, advocates for cocoa farmers' rights and environmentalists who seek to protect the last remaining rainforests in West Africa, should be more cautioned when raising demands towards TNCs from the GCCC to extend their initiatives to the local level of production. Expecting TNCs to engage at the local level of production in remote areas bears the high risk that, in the absence of solid public regulatory capacities, they push their business interests down to the supplier base. This interest, however, in a strong asymmetric constellation like the GCCC, may not necessarily be equal to what is making the target groups of their interventions better off in a long-run, and not lead to inclusive development at the local level.

Besides, in a context of younger democracies in the Global South, the major importance of improving (public) governance capacities has been broadly recognized. In the World Development Report 2017, "Governance and the Law", the World Bank (2017) highlights the key role of governance for "Ensuring Equitable Growth in Developing Countries" (World Bank 2017). The unequal distribution of power in a society is regarded to interfere with policies' effectiveness (World Bank 2017). But as it stands, when looking at power distribution, the power of TNCs and their capacity to influence and stir up local institutional environments and developmental processes is largely neglected. In the report and in broader discussions, the attention is mainly paid to national and local elites in a country. This is surprising, given the mainstream political approach of public-private partnerships and multi-stakeholder initiatives, be it for the development of rules or the implementation of sector strategies, and the thereby broadly accepted allocation of governmental and implementing power to TNCs.

In Chapter 2, the major theoretical claims from proponents of an increased political and social responsibility of TNCs have been discussed. Two key points of this theory strand are that the taking over of traditional public functions, such as the provision of social services in health and education or their engagement in self-regulation, also attributes political rights of participating in the public process of political will-formation to them (Scherer and Palazzo 2008a, p. 31). The second and similar argument is that TNCs de-facto already administer civic rights (equally through their provision of social services), particularly "where government has not yet administered citizenship rights" (Matten and Crane 2005, p. 172), what for them is mainly the case in so called "developing countries". Interestingly, the actual state of increased TNCs power in political arenas and national governments' failures to regulate private activities, is taken as an argument to further increase the influence of TNCs. Hence, in many cases, the problem becomes equally the solution. But as the discussion of the empirical case has shown, the often-stated governance gap is rather increasing through undermining national regulatory capacities instead of being closed by TNCs engagement in local domains. Thus, the claims of political CSR do not only miss a solid input legitimacy of TNCs but, as has been shown with the diverse deficits and challenges linked to their increased local engagement, even the output legitimacy.

Nevertheless, this does not mean that TNCs should completely be asked to refrain from their interventions at local levels. Indeed, they are endowed with important skills and resources and can significantly contribute to problem-solving at various scales. It is the way how their contribution is framed and regulated that is crucial. The results of the present study suggest one possible way forward: the decoupling of consent governance strategies from industrial control governance. In a 21st century world economy of global value chains, industrial governance is a substantial management tool of any lead firm. The question is about its institutional embeddedness at the various levels of supply chains. Hence, as a first step, the focus should be on the consent governance elements which allow TNCs to expand their spheres of influence into the institutional environments which potentially restrict their industrial governance, and which even lets this process appear as desirable. The discussion has given reason to embrace consent-governance' potential to mask the expansion of hegemonic movements and to block alternative change. There is therefore a need to emancipate public will-formation from elements of governance through consent. This can be achieved by continuously exposing TNCs' particular interests and the damaging effects of some of their business strategies.

8.3 Outlook: A Multiple-Scale Effort to Achieve Cocoa Sustainability and Overcome Cocoa Producers' Poverty

The present discussion has revealed several shortcomings of CSR and therewith indicates that it cannot be considered as an appropriate tool to achieve cocoa sustainability and to overcome the poverty of cocoa farmers. Moreover, due to increasing dependencies, farmers' position in the GCCC might rather deteriorate with the implementation of CSR. In that vein, Oomes and Tieben et al. (2016) concluded that policy support for TNCs' sustainability interventions increase farmers' dependency from cocoa. The downgrading of other local stakeholders in the GCCC seems to be another disregarded effect of sustainability certification implemented by TNCs. But if the often-requested amplification of TNCs' responsibilities to the local level of production tends to have such reverse effects, what could be the way forward to achieve a "real" sustainability of the GCCC which brings more inclusive forms of development?

It seems that only a multi-scale strategy can be sensitive enough to tackle the complex interplay of causes and effects, and to foster a development that really benefits cocoa farmers. Hence, a set of initiatives at different levels of the GCCC is needed that could comprise international CSR regulations, the pricing of cocoa, binding supply chain rules, dedicated national investments in remote rural areas, and emancipatory local empowerment strategies.

8.3.1   Requirements at the International Level

An international CSR regime could act as a compliment to the currently negotiated UN Binding Treaty, which provides clear rules on how to implement TNCs'
Human Rights obligations in a national state (see Section 2.3). The difference
between the two is that while the UN Binding Treaty centers around TNCs'
obligations regarding Human Rights of members within its supply chain and to
implement measures of due diligence, it lacks aspects of regulating the implementation and content of CSR interventions. While the UN Binding Treaty would
require companies to approve that any of their activities do not have negative
effects on human rights, it would not urge them to agree upon their human right
intervention with national governments and coordinate the implementation—as
for instance which areas to cover, which sector policies to respect, etc. Such a
regulation, however, could be a public governmental tool, that would tendentially
benefit all stakeholders in a sector, but if resting on voluntarism as for instance
in Ghana's Cocoa Extension Public-Private-Partnership, would not be effectively
respected.

In any case, such a potential tool for CSR regulation remains only hypothetical here. But there is a number of policy claims that have already achieved a
general consent in the cocoa sector. There is a broader acceptance that the most
urgent need is to achieve fair living incomes and wages for farmers in global
agricultural value chains (cf. Bundesministerium für wirtschaftliche Zusammenarbeit und Entwicklung n.d.). Therefore, the "Make chocolate fair"-initiative
claims for the establishment of a public-private deliberation process on how to
achieve fair living incomes wages and how to calculate the cocoa price (Bahn
and Schorling 2017). The fair living income concept that the "make chocolate
fair"-initiative bases its arguments on implies that prices should at least cover
basic needs (sufficient and healthy food, housing, drinking water and sanitation, clothing, education, health care, mobility, savings, (Bahn and Schorling
2017, author's translation) and investments associated with cocoa farming (tools,
seedlings, fertilizers and pesticides, tenancy licenses, certification, living wage for
employees, interest on loans, savings, Bahn and Schorling 2017, author's translation). The immense importance of this concept becomes clear: If cocoa farmers
were receiving a price which would allow them to finance all these livelihood
needs and production requirements, most of the voluntary engagement of TNCs
would become obsolete. No distribution of free seedlings and chemical inputs
would be needed. If there would be purchasing power for these products, the
local market would quickly evolve. One would not see children during the week

on the family farms but more farm workers. Farmers would not face such difficulties to get loans from banks because they would be able to pay them back on time. While all this sounds rather utopic at the time of writing, the major importance of the price becomes clear; as well as the fact that TNCs' voluntary interventions in these fields rather hamper locally owned forms of rural structural change. Therefore, the importance of fair prices cannot be overstated. In this context, Hütz-Adams (Hütz-Adams and Greiner 2020, p. 41) point to the fact that even if many companies are aware of the fact that considerably higher prices are needed to overcome producers' poverty, no company is willing to make the first step and pay higher prices since this would immediately translate into higher costs and reduced competitiveness for the company (Hütz-Adams and Greiner 2020, p. 41). Therefore, as it stands, here too, some regulatory measures at national and international scales seem required to exert enough pressure for the needed change. In Germany, the third biggest importing country of cocoa products, for instance, the Federal Ministry for Economic Cooperation and Development launched a multi-stakeholder initiative (MSI) for sustainable agricultural supply chains (Initiative für nachhaltige Agrarlieferketten, INA) which has a distinct focus on attaining fair living incomes and wages across sectors and supply chains (Bundesministerium für wirtschaftliche Zusammenarbeit und Entwicklung 2019). It can be hoped that in this important MSI, TNCs do not trade on their dominant positions and that tools and objectives do not become diluted.[4]

Besides, in Germany, a civil society initiative for a fair German supply chain act (Lieferkettengesetz), recently gained broader political support, especially by the German foreign and the development ministers, and might still be approved in the present legislature period. The five main claims of this initiative concern the fields of due diligence along supply chains, control and sanctioning mechanisms for a public agency, the recognition of the link between the destruction of the environment and human rights violations, civil law liability, and companies concerned should also include smaller and middle-size firms (Initiative Lieferkettengesetz n.d.). With this set of claims, the respect of human rights along global value chains by the private sector is supposed to be enforced. While such a legal act is a crucial step forward in direction to tame profit orientation, and while it provides important measures to sanction specific cases of human rights violations,

---

[4] Three important civil society organizations in Germany, the Network for Corporate Accountability, together with the Forum Human Rights (Forum Menschenrechte e.V.) and VENRO—the umbrella association of German NGOs in development cooperation) have listed eight requirements for efficient MSI: sound preparation, binding principles, binding targets, governance structures, transparency, impact measurement, complaint mechanisms and sanctions, enabling framework, Heydenreich and Guhr 2020.

the question remains open, to which extent it will also be able to legally address one of the key structural factors of human rights violation: far too low incomes and wages at the bottom of global value chains.

8.3.2 The Need for Increased National Efforts

At the same time, while increasing the price for cocoa beans, a comprehensive set of national investment to foster local infrastructure would be needed. In Section 6.5. on the characteristics of the research area, the huge infrastructural deficits of the Western North region in Ghana were described. Main deficiencies concern the road network, making remote areas almost inaccessible, weak electricity networks to which some of the remotest communities are still not even connected, lack of clean water and sanitation facilities and a low coverage with secondary schools and clinics. The bad road network particularly isolates dwellers of remote communities from ongoing developmental processes in the area and makes it harder and more costly for them to contribute to local markets which increases their opportunity costs of farm diversification. Dedicated public investments in all these key infrastructure areas in the rural margins would trigger multiple positive dynamics and contribute to the expansion of local input markets and access to many other services. Simultaneously, farmers' human and social capitals would be vitalized by better access to education, and health and sanitation services. Nevertheless, as simple as this sounds, it is as complex in its realization. Next to issues of corruption, political clientelism and inefficiencies in service delivery, the main hinderance for such broader infrastructure investment is the lack of public budget in most countries of the Global South. Many countries in the Global South are confronted with high and continuously increasing shares of their government revenues to be used to pay back their foreign debts, and which therefore cannot be used for health or social infrastructure expenditures, for instance (Falk 2020). Because of these national development-hampering debt burdens and an impending debt crisis, one of the main claims of the United Nations Conference on Trade and Development (UNCTAD) is the "Rechannelling of external liabilities (debt) into productivity-enhancing domestic investment, building "real patient capital" at home to support and enable domestic structural transformation" (United Nations Conference on Trade and Development 2020). The allocation of investment to the expansion of rural infrastructure in order to facilitate local economic diversification and to improve educational levels of rural dwellers goes in line with this claim.

The establishment and expansion of alternative markets in frontier areas is an important aspect of fostering economic diversification and of reducing farmers' dependencies from only one cash crop as income source. The importance of income diversification for the improvement of rural livelihoods has received much attention in rural development theory. Generally spoken, diversification can be regarded as a form of self-assurance. Most important "push-factors" are the reduction of risks, diminishing returns to labor and land or the coping with external shocks. But there is also the households' strategic project to find best constellations and complementarities between activities and to maximize returns from them, the "pull-factors" (Barrett et al. 2001, p. 315f.). The objective to reduce households' reliance from only one income source can be realized through two main approaches of rural income diversification, that is on-farm and off-farm diversification. While on-farm diversification comprises strategies of mixed cropping or intercropping or the interlocking of crop and livestock production activities (Olayiwola 2013, p. 28), off-farm diversification includes off-farm wage work in agriculture, wage work in non-farm activities or rural non-farm self-employment (Ellis 2000, p. 292). However, off-farm diversification activities show a higher correlation with wealth accumulation than on-farm ones, but are more difficult to achieve for the very poor which only have a weak asset endowment. Therefore, poverty policies should concentrate on improving the endowment of assets of the rural poor which can be contributed to their diversification strategies (Barrett et al. 2001, p. 316).

In the case of support of on-farm diversification, it does not only imply the production of other agri-food products, such as rubber or cashew, timber trees or food crops for the local markets. But it would also require all stakeholders to finally take existing local knowledge and innovation capacities into account when designing development interventions. The planting of cashew trees, for instance, on cocoa farms improves the agroecological performance of the farms and provides an additional income source. A study conducted in the Ivorian cocoa sector has carved out how the planting of cashew trees contributes to a local agroecological transition driven by innovative cocoa farmers—and not by external advice: In a context of a declining forest rent and a stagnation of cocoa prices, given the stronger resistance of cashew trees to droughts, the trees' minor need of chemical inputs, and their function to improve land tenure security in local tenure systems, farmers have recently started to increase the number of cashew trees on their cocoa farms (Ruf et al. 2019a). This locally-made innovation is one example of farmers' capacities to take their farming decisions based on their own knowledge and interests. In fact, the importance of farmers' traditional knowledge for their adaptation to changing climatic conditions and the end of the forest rent seems to be largely underestimated in present extension approaches. In their field work, Sanial and Ruf (2018) have observed how, in the course of the current sustainability certification projects, farmers are often "perceived as uninformed, unprofessional, incapable of innovating and adopting rational economic behaviour" (Sanial and Ruf 2018, p. 161). Therewith, in the top-down form as currently practiced, certification with its extension services embodies a paternalistic approach which tends to ignore local knowledge and innovation capacities of the targeted farmers, such as their already existing practices in agroforestry. This, in its final essence, represents a continuity of agronomic practices applied during the colonial period (Sanial and Ruf 2018).

This observation mirrors a more general reflection on development dynamics in rural Africa. Charlery de la Masselière (2005) has described the process of agricultural modernization in Africa as a continual form of domination through the project of technization and the introduction of a productivist logic. While the main interest of the colonial rulers was the mobilization of African peasants' workforce for the desired cash crops, they largely ignored the existing customary practices and institutions as well as the socio-economic relations in the context of agricultural labor and land tenure practices. The project of modernization created a structural dualism between urban and rural territories, where the first hosts a civil society oriented towards Western lifestyles, and the latter a peasantry which mainly still bases on customary practices but seeks to integrate into the urban-centered process of modernization, as most visible in the extensive rural exodus. Congruently to this divide, rural workforce and land are extracted from its original meanings and designated to serve the markets of the urban centers—within the country and abroad. Therefore, it appears that many African rural societies are confronted with the need to find their way with this structural "unfinished modernization" (Charlery de la Masselière 2005, p. 46, author's translation). In this situation, many rural dwellers are torn between customary practices and the decreed production rationality with its tools for intensification which remained largely extraneous to local production meanings and values.

It seems to be important to take this underlying structural discord in many rural areas in Africa more precisely into account when pondering on the future of a cocoa sustainability in West Africa and on a developmental process, which pre-eminently improves farmers living conditions. One crucial first step could be the appreciation—and not stigmatization—of the remaining customary practices. In a next step, farmers, in an emancipated exchange with multiple stakeholders and facilitated by a more efficient public extension system, could weigh which practices fit best to their needs and interests and jointly explore combinations of techniques with stakeholders. For instance, in her study on cocoa farmer-owned innovations, Sanial (2019) carves out how cocoa farmers pro-actively respond to increasing land shortage under a "post-forest" condition, hence, the cocoa frontier being at its limits. Farmers own adaptation practices tend to expand agroforest schemes in their farms—instead of blindly following the intensification advices by certification schemes. Such a farmer-driven local innovation process is one of the key ideas of the agroecology and food sovereignty movements.

8.3.3   The Gains of Local Self-Organization

Food Sovereignty as a global political peasant movement dates back to the 1980 s and is a grassroots response to the many neo-liberal reforms in the era of the Structural Adjustment Programs. The broad international peasants' coalition, La Via Campesina, introduced the concept of a Food Sovereignty at the World Food Summit in 1996 and used it to advocate for the protection and enhancement of domestic farming systems (McMichael and Porter 2018). From then onwards, the growing globally organized peasant movement has made use of the concept for different claims, such as demands for land rights, the preservation of traditional seed systems or the creation of training schools for agro-ecological farming methods (McMichael and Porter 2018, p. 209). In 2007, over 600 civil society representatives came together in the Malian countryside and hold the International Forum on Food Sovereignty, "Nyéléni 2007" (Ratjen 2008). The Nyéléni Declaration is the major outcome. Some of the major claims which were stated concern, among many other aspects, the self-determination of peoples' own food producing systems, the ability to live with dignity and to earn a living wage for labor, the conservation and rehabilitation of rural environments or food traditions, the recognition and respect of peoples' diversity of traditional knowledge or the way how they organize and express themselves (La Via Campesina 2007). The international movement is mainly represented by La Via Campesina. In the Americas, all countries (except Guyana and Surinam) have member groups in the umbrella NGO but in Africa it is only about the half of the countries. In Ghana, the organization ECASARD—Ecumenical Association for Sustainable Agriculture and Rural Development—is a member of the umbrella NGO and a Food Sovereignty group also exists[5].

Yet, while the mentioned claims and the ideas of this global grassroots movement definitively concern the interests of poor Ghanaian cocoa farmers, the idea

---

[5] See http://foodsovereigntyghana.org/about-us/ (last accessed on 04.10.2020).

to organize themselves and articulate their needs is not widespread in cocoa production frontiers in Ghana. The reasons for this are likely numerous, and may equally rest in the "unfinished modernization" project the difficult access to information diversity, the fixed price for cocoa beans, or the poverty trap of one-sided integration in a GVC and the consent strategies through a top-down "empowerment" exerted by TNCs. In order to support bottom-up movements and farmers' self-organization, a focus should be on facilitating formal education, improve communication infrastructure, valuing local innovations, and encouraging the exchange between farmers groups.

Conclusions

9

This study was primarily concerned with unpacking the governmental and power implications of an enlarged voluntary social responsibility of TNCs in global sustainability governance, more precisely through sustainability interventions in their supply chains. For this discussion, the study has taken the example of the Global Cocoa-Chocolate Chain (GCCC) and the implementation of an UTZ cocoa sustainability certification project in Ghana's most productive cocoa growing area by one of the biggest transnational agricultural companies in the GCCC.

The starting point of this study was a situation where TNCs in the GCCC are faced with an increase in sustainability risks for local cocoa production in West Africa, the major cocoa producing area in the world from which they are strongly dependent. With a detailed study of the composition and functioning of the GCCC and the cocoa sector in Ghana, this study has provided a systematic overview of current economic, environmental, and social challenges for cocoa production and their entanglement with institutional environments. From a TNC perspective, whose main particular interest lies in the assurance of a lasting flow of high-quality cocoa beans, the main sustainability risks for cocoa production include low levels of cocoa farmers' productivity, cocoa farmers' poverty, a poor level of sector modernization, and ongoing dynamics of cocoa frontier diversification. Moreover, the major challenge for TNCs' from the GCCC to control sustainable production flows in their supply chain lies in the nature of small-scale cocoa production in West Africa: a highly fragmented production base situated on remote rural areas which is difficult to access and streamline.

Hence, the main epistemological interest of the empirical study was to identify how and by which means TNCs foster their governance aspirations through the implementation of their CSR interventions. The guiding questions of the study were how these interventions change the local setting of Ghana's cocoa sector

F. Ollendorf, *The Transformative Potential of Corporate Social Responsibility in the Global Cocoa-Chocolate Chain*, (Re-)konstruktionen – Internationale und Globale Studien, https://doi.org/10.1007/978-3-658-43668-1_9

and whether this helps the implementing TNCs to increase their control over local production processes and change the institutional environment in a way that facilitates their interests. In this context, the perceptions of cocoa farmers towards their cocoa farming and existing livelihoods also played an important role. These insights in local sector dynamics stemming from transnational CSR provided the basis for a discussion of TNC-driven sustainability governance and the implications for the target areas as well as power constellations in the GCCC. Such an analysis is an important contribution to global justice discussions and helps to take an informed standpoint on whether to hand over more responsibilities to TNCs and if so, how this should be best institutionalized.

In order to comprehensively capture the governance dynamics linked to transnational CSR, the study laid out a conceptual framework which allows to approach TNCs' CSR activities as tools for their pursuit of control in their global value chains and the needed consent for it. A consent and control framework was developed which combines the elements of Global Value Chain Analysis with neo-Gramscian aspects of global governance theory. This nexus materialized in a set of axes, including actors, geographical scope, industrial governance, input-output structure, institutional environment, means of action, perceptions, scales, and topic of consent creation, which together grasp the implementation process and outcomes of consent and control-governance efforts. In the methods section, these axes were operationalized to orient the field work and to guide the process of data acquisition, analysis, and finally the discussion.

The concluding chapter is organized as follows: In a first step, the key research findings of the study are summarized again. In a second step, the challenges to approach the field and the limitations of the findings and their transfer to a broader level are explained. Finally, in the last part of the chapter and hence, this study, the strengths of the findings and their possible implications for the discussion of sustainability governance and the roles of different actors are shown.

9.1   Unpacking Ambitions, Tools and Effects of Transnational CSR

The institutional changes linked to the ensemble of several simultaneously ongoing sustainability interventions by large transnational cocoa processors and some chocolate manufacturers have been described in the previous chapters. The empirical study revealed three main areas of sectoral change linked to sustainability certification projects currently ongoing in Ghana: sector cooperation, sector organization, and marketing relations. Regarding these three areas, tendencies seem rather clear whereas the real benefits of CSR for cocoa farmers remain controversial.

In the course of the implementation of TNC-led sustainability certification, new forms of sector cooperation emerge. This is most obvious when looking at the new alliances between TNCs with either LBCs or NGOs or both. These alliances are stronger forms of cooperation than usual partnerships, because the stakeholders merge to one joint player (in the presented case the project group) who benefits from the strengths of each member and strives for the same goals and interests. Indeed, such alliances are very efficient in action, pooling the different assets of each member. In the presented case, the lead firm of the project was strongly benefitting from this constellation as long as it needed. Next to these close alliances, partnerships with the public agency in charge of cocoa, the Ghana COCOBOD, also evolved in the course of the certification project. But this partnership does not go beyond minor aspects of local implementation and importantly does not involve planning of the intervention or harmonizing it with other sustainability activities in the sector. It therefore appears that the partnership approach is currently practiced on a "pick and choose" basis in accordance with the main interests of TNCs.

Other important changes concern the organization of the sector which is also affected by the introduction of new forms of extension services and new input delivery schemes by the transnational private sector. While COCOBOD is running its public extension system and some input delivery support programs, the implementation of the private structures creates a parallel system which is very little coordinated with the existing public services. The private extension system, based on the "train the trainers"-approach, is more efficient and appropriate to the challenge of reaching as many cocoa farmers in remote communities as possible than the public one. But, as it stands, extension activities in the course of certification projects, are of exclusive nature and not open for all farmers of a community but only to the ones targeted by the intervention. At the same time, public extension agents go less to those communities who benefit from a private intervention what implies a potential to raise the extension gap for the non-certification-targeted farmers in project communities. A similar pattern can be observed for the access to improved farming inputs. The tendential privatization of extension services can be regarded as an ambivalent change of the local sector organization. It is a positive development that targeted farmers are benefitting from improved access to quality inputs and different forms of trainings. But at the same time, given that all services for each project ongoing respectively are provided or coordinated by the same powerful TNC, the dependencies of the farmers, which are now bound to that company by contract, drastically increase. Furthermore, such developments and linked improvements are limited to the participating farmers while the rest of the community remains excluded from it. With this, new forms of inequality at the local level are created.

Yet another major change at the local institutional level occurs at the system of internal marketing. Certification, attracting many farmers to sell their beans to the implementing LBC, became a strong tool for marketing competition between LBCs, which due to the fixed price for cocoa, do not compete on a price basis in Ghana, but through other incentives for farmers. Hence, the premium and the distribution of some minor material inputs, such as personal protection equipment, and the training programs on farming practices, became important tools for competition. This study has argued that this reduces the competitiveness of local LBCs which mostly lack the resource endowment to run the complex Internal Management System (IMS) needed to implement a certification project.

With the tendential concentration of large LBCs, the internal marketing system also changes. In fact, a new local supply chain emerged. Certified beans undergo a strict quality check conducted by IMS staff in addition to the traditional public quality checks which makes the latter almost obsolete. Certified beans are stored and transported separately and gain a privileged access to depots next to the port what constitutes a substantial financial advantage for LBCs with certification projects. Moreover, effects of the growing importance of certification on the external marketing system were reported by interviewees too. Since COCOBOD is not involved in the negotiations on expected volumes of certified beans, its role in the certified supply chain is less important than in the conventional chain. It rather has to adapt and integrate the new structure into its own system. Arrangements are made directly between transnational processors and LBCs which COCOBOD only has to assign. Thereby, COCOBOD is concerned with keeping the good marketing relations with its international buyers. However, the change is becoming even more profound since due to their sustainability strategies, the leading transnational processors (Cargill, Barry Callebaut, Olam, Touton) have created their own LBCs and therewith expanded their backwards integration. In Ghana, the local presence of TNCs was cut-off after independence of the country, but this policy seems to be reverted by the current trend.

It appears that, linked to the implementation of transnational CSR strategies, which is becoming ever more pro-active in the past few years as shown with the example of studied project, COCOBOD's role in the sector is changing too. First of all, COCOBOD does not have a real stake in the expounded sector changes but has to accommodate them. Therewith, one could argue, its sovereignty in the sector is subverted and a new constellation in which it has to share responsibilities with the transnational private sector emerged. For instance, while TNCs just went on their own to establish their projects, ownership in decision-making or policy implementation of COCOBOD became affected. The creation of the Project Coordination Unit by COCOBOD is one indicator for the board's will to gain back control over these developments in order to harmonize them with its own policies. Hence, in this study, it has been argued that the implementation of sustainability certification schemes by the transnational private sector does not come without the risk of forfeiting public regulatory capacities, particularly in remote areas with weaker governmental structures.

Getting back to the point of departure of the present study, the underlying hypotheses, backed by the consent and control framework, assume the need for a two-sided governance strategy of TNCs which seek to modernize the cocoa sector and professionalize cocoa farmers in order to assure the sustainability of the flow of good quality beans while keeping the status quo of unfair distributional patterns in the chain. One side, the control side, implies industrial means of governance and seeks to increase control over local production processes. It is closely related to conducting actions which transform the institutional environment of the chain in a way that more control can be exerted. The study presented important tools of industrial control governance, such as the implementation of contracts for farmers participating in the project or the establishment of monitoring by large documentation efforts and sanctioning mechanisms in the IMS. The above described institutional transformations can be regarded as "successful" outcomes of governance through control measures. The second side of governance concerns perceptions of sector stakeholders, especially of cocoa farmers, who need to be convinced to remain in the sector instead of investing in other livelihood strategies. This governance through consent-dimension is more subtle than the control dimension and more difficult to capture. A neo-Gramscian perspective on governance opens the view for such subliminal governance aspirations by analyzing efforts to build alliances and foster partnerships as governance tools which serve to spread consent for the own project. Similarly, the making of compromises and giving concessions by TNCs can be regarded as strategies which seek to incorporate critique on the status quo and prevent uprising by subalterns in the chain. In this light, the present study has discussed the above presented sector transformations such as building alliances and entering partnerships as means of consent creation. Further, the distribution of the premium and other materials to farmers as well as the improved access to training and extension services, have been presented as concessions which aim to keep the farmers in the sector and prevent their reorientation towards alternative livelihood strategies. But in contrast to the control dimension, the consent governance seems to be less successful. Even if participating in a certification project, farmers are still trapped in poverty and struggle to achieve their basic needs. Hence, also in a sustainability certified cocoa chain, there is no fairness yet. For many of the interviewed farmers, the production of cocoa therefore is still only regarded as a strategy to enable them or their children to grow into other business activities. However, this, from the consent and control framework perspective, is not even surprising. CSR as a governance tool seeks to prevent the status quo of distributional patterns, and therefore is not meant to tackle underlying structural reasons for poverty but only to blur them. Therefore, when being conducted in a voluntary form and without regulation, CSR can bring only short-term improvements to a sector. But thereby, it possibly reinforces asymmetric dependencies, in the studied case of small-scale farmers from transnational processors, and likely hinders a more profound sector change which would benefit society in a broader way.

For these reasons, the present study suggests that the Ghanaian cocoa sector may reach a crossroads in the near future. One possible path leads to a strong role of TNCs at the local level and a wide modernization of the cocoa sector largely maneuvered by TNCs with a smaller number of highly professionalized farmers remaining in the sector. A large number of cocoa farmers would have dropped out of the sector in this scenario. The second path could be an emancipated and more diversified rural society while the public agency guards its steering role in the cocoa sector. Intense public investment, made possible by programs of international debt relief, may ease a process of frontier diversification and enable higher levels of education of rural dwellers. Due to the existence of alternatives, cocoa farmers became less dependent from cocoa and make real choices regarding their future. In this constellation, the local cocoa market would not be controlled by a few TNCs but mainly run by local companies. If cocoa prices grew and farmers would achieve fair living incomes, certification in its current form would become obsolete and all farmers could benefit from better prices.

Obviously, these two scenarios are speculative and draw two idealistic pictures of opposing sector dynamics. By drawing these extremes, the study does not seek to model or forecast possible sector developments. Rather, the brief sketch of these two opposing ways forward invites to open the view for the importance to reflect on long-term effects of sector interventions and on the possible structural changes which might stem from them.

9.2   The Challenge to Open the Box—and How to Look in it

The achievement of a sound empirical evidence on industrial governance processes is challenging since here, particularly, information is involved that often contains competitive information and hence is regarded as confidential. At the same time, even if a great number of farmers were interviewed and the author spent several months in the research area, grasping the complete picture of what matters for cocoa farmers and how they really look at ongoing sector dynamics remains difficult if not impossible under the given conditions. Furthermore, sectoral dynamics are fast, and as a researcher not constantly located close to them, the risk that findings become outdated is high.

When talking about *effects*, the study sought to remain cautious and explicitly does not use the term "impacts". There is not enough empirical ground of the study to state clear cause and effect relations. Rather, tendencies were brought to light which indicate the existence of empirical effects. Further, the objective was not only to set out causal relations, but to illuminate their underlying *mechanisms*. Thus, the presented results are only able to capture tendencies and the subsequent discussions have to be read in this vein. Because of this, and because of the sector specifics, as already stated in several parts of this study, an immediate transfer of the results to other countries or other value chains is not possible. Nevertheless, the discussion enables to capture important features of TNC-led sustainability interventions in a context of weaker local regulatory capacities. During literature review, some studies with similar findings and which trace similar dynamics linked to sustainability certification could be identified. They point to similar dynamics and therefore seem to confirm that the presented results are also of relevance for other constellations. Still, another challenge to open the box of the transformative potential of CSR was the almost inexistence of studies which look at more structural local implications of transnational CSR and which empirically approach certification from a power-sensitive angle.

It is broadly acknowledged that researchers are not free from their personal standpoints and biased to approach their research topic from the lenses through which they see the world. While in social science research, it is especially difficult to overcome this bias, it is ever more important to transparently expose this bias. In the case of the present study, the author reflects to be personally guided by an institutional notion of justice as set out in the introductory chapter. This means that her thoughts are based on an understanding of justice, which concedes equal opportunities to access resources and positions to all mankind and in which institutions have to guarantee the fulfillment of human rights. This makes the study biased towards discovering structural power asymmetries in social relations and approaching them as institutional challenges which need to be overcome instead of seeing them as windows of opportunity—as defenders of political CSR would do. While this normative position of the author has likely influenced the research design, given the transparent research process, the results are able to speak for themselves and show a current dynamic of TNCs' power expansion that needs further attention in future.

9.3   Possible Future Forms of CSR and Implications for Research and Practice

While at the point of writing it might still sound a bit fictional, the future of CSR or sustainable supply chain management will likely lie in the digitization of as many aspects as possible in supply chains. *Methods of precision agriculture in a fragmented cocoa production base like in Ghana, would that ever work out?*—CEOs of TNCs being active in many supply chains might ponder and seek to transfer methods they apply in other production contexts to the GCCC. In fact, during the field work in 2015 and 2017, new technologies were already being implemented. A simple starter, for instance, was the implementation of the mobile phones outreach program COCOLINK in a partnership program between WCF and COCOBOD (World Cocoa Foundation 2012), but efforts to digitally spread extension messages will definitely not stop there (cf. Ollendorf 2017). Moreover, the introduction of farmer ID cards, where farmers' production data are gathered and transferred to data management tool is about to come and indicates the development of blockchain capacities. Likewise, the increase in full traceability schemes, the gathering of soil and climatic data via GPS or possibly drones, and the improved digital monitoring of harvest and diseases all contribute to the optimization of inputs and farming techniques, hence to precision farming.

All these technologies only very recently gain importance in the cocoa sector, and hence, there is currently little knowledge on how these advances are implemented and, more importantly, how the knowledge gathered with these techniques will then be transferred to the farming level. Will the possession of so many farming data help TNCs to increase local production capacities and if yes, how? Moreover, how will this again shape the local context of small-holder production?

In any case it seems very likely that modernization efforts by both TNCs and public agencies in the course of their sustainability management will follow the global trend in agricultural production and seek to apply high-tech measures in the field. While these attempts will face a local reality of low levels of infrastructure, the combination of all efforts will surely impact the local sector functioning in one way or another.

Considering this future perspective, the digitization of the GCCC would add to the already observed transformative dynamics triggered by CSR. Since the societal implications of such a project of sector modernization and increased monitoring are potentially manifold and sustainability efforts are amplifying fast, it is an important future task to research transformative processes linked to CSR and sustainability governance. Thereby, a special attention should also be paid to farmers' emancipatory potential: How and why do farmers decide to include new technologies in their own practices or not? Is there a potential that some new technologies become a tool for them to improve their position in the chain, for instance by organizing themselves via social media or developing some ways to make strategic use of their farming data? This, however, is a hard piece of work, what becomes more apparent when looking at the many frontiers of capitalism which put small-holder farmers throughout Africa and elsewhere under pressure (cf. Metzger et al. 2016).

Finally, the key message of the study is that, when urging TNCs to assume more voluntary responsibility in their supply chains, the structural implications of such an expanded responsibility need to be carefully assessed and taken into account. Even if it seems tempting that the strongest actors in a chain should also assume more responsibility and implement some measures to improve local production conditions, it has to be considered that this might come along with side-effects that could have even reverse effects on the targeted areas. Therefore, more holistic solutions to value chain grievances are urgently needed which enable a more participatory development and eradicate systemic poverty at the bottom ends of global value chains.

References

Abbott, Kenneth W. (2012): The Transnational Regime Complex for Climate Change. Environment and Planning C: Government and Policy, 30(4), 571–590. In *Environ Plann C Gov Policy* 30 (4), pp. 571–590. DOI: https://doi.org/10.1068/C11127.

Abbott, Kenneth W.; Snidal, Duncan (2010): International regulation without international government: Improving IO performance through orchestration. In *Rev Int Organ* 5 (3), pp. 315–344. DOI: https://doi.org/10.1007/s11558-010-9092-3.

Abdulsamad, Ajmal; Frederick, Stacey; Guinn, Andrew; Gereffi, Gary (2015): Pro-Poor Development and Power Asymmetries in Global Value Chains. Duke University Center on Globalisation, Governance and Competitiveness. Available online at http://www.cggc.duke.edu/pdfs/Pro-PoorDevelopment_and_PowerAsymmetries_inGlobalValueChains_Final.pdf, checked on 6/13/2016.

ACET (2014): The Cocoa Agri-Processing Opportunity in Africa. Background paper for the 2014 African Transformation Report. With assistance of Dalberg Global Development Advisors. African Centre for Economic Transition. Available online at http://acetforafrica.org/publication/the-cocoa-agroprocessing-opportunity-in-africa/, checked on 11/4/2017.

Acquier, Aurélien; Daudigeos, Thibault; Valiorgue, Bertrand (2011): Responsabiliser les chaînes de valeur éclatées. In *Revue francaise de gestion* N° 215 (6), pp. 167–183. Available online at https://www.cairn.info/revue-francaise-de-gestion-2011-6-page-167.htm.

Agergaard, Jytte; Fold, Niels; Gough, Katherine V. (2010): Introduction. In Jytte Agergaard, Niels Fold, Katherine V. Gough (Eds.): Rural-urban dynamics. Livelihoods, mobility and markets in African and Asian frontiers. London, New York: Routledge (Routledge studies in human geography, 29), pp. 1–8.

Almeida, Karin; Azemedo, Simone; Wodarg, Frithjof (2012): CSR Instruments: A Systematic Overview Evaluation for the German CSR Forum and Bertelsmann Foundation. Edited by Hertie School of Governance. Available online at https://opus4.kobv.de/opus4-hsog/frontdoor/index/index/docId/2041, checked on 9/10/2020.

Amegashie-Duvon, Edem (2014): Understanding the perceptions of traceability in the cocoa supply chain: the case of Ghana. Doctoral thesis. Newcastle University, Newcastle. Newcastle University Business School. Available online at https://theses.ncl.ac.uk/dspace/bitstream/10443/2623/1/Amegashie-Duvon,%20E.%2014.pdf, checked on 9/28/2016.

Ametepeh, Emmanuel (2017): Forest Transition Deficiency Syndrome. The case of forest communities in the high forest zone in Ghana. Dissertation. Justus Liebig University, Giessen. Institute of Political Science.

Ameyaw, K.; Ettl, Gregory J.; Leissle, Kristy; Anim-Kwapong, Gilbert J. (2018): Cocoa and Climate Change: Insights from Smallholder Cocoa Producers in Ghana Regarding Challenges in Implementing Climate Change Mitigation Strategies. In *Forests*.

Amoah, J. E. K. (1998): Marketing of Ghana cocoa, 1885–1992. Accra-North, Ghana: Jemre Enterprises (Cocoa outline series, no. 2).

Anlauf, Axel; Schmalz, Stefan (2019): Globalisierung und Ungleichheit. In Karin Fischer, Margarete Grandner (Eds.): Globale Ungleichheit. Über Zusammenhänge von Kolonialismus, Arbeitsverhältnissen und Naturverbrauch. Wien, Berlin: Mandelbaum Verlag.

Ansah Go; Fnk. Ofori; Lawrencia Pokuah Siaw (2018): Rethinking Ghanas Cocoa Quality: The Stake of License Buying Companies (LBCs) in Ghana. In *Journal of Agricultural Science and Food Research* 9 (2).

Anti-Slavery International (2004): The Cocoa Industry in West Africa: A history of exploitation. Available online at http://www.antislavery.org/wp-content/uploads/2017/01/1_cocoa_report_2004.pdf, checked on 11/4/2017.

Anyidoho, Nana Akua; Leavy, Jennifer; Asenso-Okyere, Kwadwo (2012): Perceptions and Aspirations: A Case Study of Young People in Ghana's Cocoa Sector. 43rd ed. (IDS Bulletin, 6).

Aoudji, Augustin K. N.; Avocevou-Ayisso, Carolle; Adégbidi, Anselme; Gbénou, Cassien; Lebailly, Philippe (2017): Upgrading opportunities in agricultural value chains: Lessons from the analysis of the consumption of processed pineapple products in southern Benin. In *African Journal of Science, Technology, Innovation and Development* 9 (6), pp. 729–737. DOI: https://doi.org/10.1080/20421338.2016.1163472.

Araujo Bonjean, Catherine; Brun, Jean-Francois (2016): Concentration and Price Transmission in the Cocoa-Chocolate Chain. In Mara P. Squicciarini, Johan F. M. Swinnen (Eds.): The economics of chocolate. Oxford: Oxford University Press, pp. 339–362.

Armando, Eduardo; Azevedo, Ana Claudia; Fischmann, Adalberto Americo; Pereira, Cristina Espinheira Costa (2016): Business strategy and upgrading in global value chains: a multiple case study in Information Technology firms of Brazilian origin. In *RAI Revista de Administração e Inovação* 13 (1), pp. 39–47. DOI: https://doi.org/10.1016/j.rai.2016.01.002.

Arrighi, Giovanni (1990): The developmentalist illusion: a reconceptualization of the semiperiphery. In W. Martin (Ed.): Semiperipheral States in the World-Economy. New York: Greenwood Press, pp. 11–42.

Asare, Richard; Afari-Sefa, Victor; Osei-Owusu, Yaw; Pabi, Opoku (2014): Cocoa agroforestry for increasing forest connectivity in a fragmented landscape in Ghana. In Agroforest Syst 88 (6), pp. 1143–1156. DOI: 10.1007/s10457-014-9688-3.

Austin, Gareth (1996): Mode of Production or Mode of Cultivation: Explaining the Failure of European Cocoa Planters in Competition with African Farmers in Colonial Ghana. In W. G. Clarence-Smith (Ed.): Cocoa pioneer fronts since 1800. The role of smallholders, planters, and merchants. New York: St. Martin's Press, pp. 154–175.

Baah, Francis (2008): Harnessing farmer associations as channels for enhanced anagement of cocoa holdings in Ghana. In *Scientific Research and Essay* 3 (9), pp. 395–400.

Baah, Francis; Anchirinah, V.; Badu-Yeboah, A. (2009): Perceptions of extension agents on information exchange with cocoa farmers in the Eastern region of Ghana. Available online at http://www.academicjournals.org/SRE, checked on 9/12/2020.

Bahn, Evelyn; Schorling, Johannes (2017): Die bittere Wahrheit über Schokolade. Edited by INKOTA-netzwerk (Infoblätter, 1). Available online at https://de.makechocolatefair.org/ material, checked on 9/12/2020.

Bair, Jennifer (2005): Global Capitalism and Commodity Chains. Looking Back, Going Forward. In *Competition & Change* 9 (2), pp. 153–180. DOI: https://doi.org/10.1179/102452 905X45382.

Bair, Jennifer (2008): Analysing global economic organization. Embedded networks and global chains compared. In *Economy & Soc.* 37 (3), pp. 339–364. DOI: https://doi.org/ 10.1080/03085140802172664.

Bair, Jennifer (Ed.) (2009): Frontiers of commodity chain research. Stanford, Calif.: Stanford University Press.

Bair, Jennifer; Palpacuer, Florence (2015): CSR beyond the corporation: contested governance in global value chains. In *Global Networks* 15 (supplemental issue), pp. 1–19, checked on 5/1/2016.

Bair, Jennifer; Werner, Marion (2011): Commodity Chains and the Uneven Geographies of Global Capitalism. A Disarticulations Perspective. In *Environ Plan A* 43 (5), pp. 988–997. DOI: https://doi.org/10.1068/a43505.

Banerjee, Subhabrata Bobby (2011): Corporate social responsibility. The good, the bad and the ugly. Cheltenham: Edward Elgar.

Barbier, Edward B. (2005): Frontier Expansion and Economic Development. In *Contemporary Economic Policy* 23 (2), pp. 286–303. DOI: https://doi.org/10.1093/cep/byi022.

Barfuss, Thomas; Jehle, Peter (2017): Antonio Gramsci zur Einführung. 2., ergänzte Aufl. Hamburg: Junius (Zur Einführung).

Barral, Stephanie; Ruf, Francois (2012): Plantations industrielles ou familiales ? Regards croisés sur la production d'huile de palme et de cacao en Indonésie et au Ghana. 2012/ 3 – N° 62 pages 75 à 93. In *Presses de Science Po* (62), pp. 75–93. Available online at http://www.cairn.info/revue-autrepart-2012-3-page-75.htm, checked on 9/10/2020.

Barrett, C.B; Reardon, T.; Webb, P. (2001): Nonfarm income diversification and household livelihood strategies in rural Africa: concepts, dynamics, and policy implications. In *Food Policy* 26 (4), pp. 315–331. DOI: https://doi.org/10.1016/S0306-9192(01)00014-8.

Barrientos, Stephanie (2011): Beyond Fair Trade: Why are Mainstream Chocolate Companies Pursuing Social and Economic Sustainability in Cocoa Sourcing? Paper to ILO/IFC Better Work Conference (ILO/IFC Better Work Conference).

Barrientos, Stephanie (2016): Promoting Gender Equality in the Cocoa-Chocolate Value Chain: Opportunities and Challenges in Ghana. Edited by University of Manchester. Global Development Institute (Working Paper Series).

Barrientos, Stephanie; Gereffi, Gary; Rossi, Arianna (2010): Economic and social upgrading in global production networks. Developing a framework for analysis. Manchester (Capturing the gains working paper, 03).

Barry Callebaut (2020): About Us. Available online at https://www.barry-callebaut.com/de-DE/group/about-us, checked on 10/1/2020.

Bartley, Tim (2007): Institutional Emergence in an Era of Globalization: The Rise of Transnational Private Regulation of Labor and Environmental Conditions. In *American Journal of Sociology* 113 (2), pp. 297–351. DOI: https://doi.org/10.1086/518871.

Beckman, Björn (1976): Organising the farmers : cocoa politics and national development in Ghana. Uppsala: Uppsala Offset Centre AB.

Beisheim, Marianna; Ellersiek, Anne (2017): Partnerschaften im Dienst der Agenda 2030 für nachhaltige Entwicklung. Transformativ, inklusiv und verantwortlich? Stiftung Wissenschaft und Politik (SWP-Studie 2017, S 22). Available online at https://www.swp-ber lin.org/publikation/partnerschaften-im-dienst-der-agenda-2030-fuer-nachhaltige-entwic klung/, checked on 9/10/2020.

Bieling, Hans-Jürgen (2011a): Discussion Paper. Edited by Zentrum für Ökonomische und Soziologische Studien. Universität Hamburg. Hamburg.

Bieling, Hans-Jürgen (2011b): Varieties of Capitalism, Regulationstheorie und neogramscianische IPÖ – komplementäre oder gegensätzliche Perspektiven des globalisierten Kapitalismus? In Zentrum für Ökonomische und Soziologische Studien (Ed.): Discussion Paper. Universität Hamburg. Hamburg. Available online at https://www.wiso.uni-hamburg.de/fachbereich-sozoek/professuren/heise/zoess/publikationen/dp23.pdf.

Bieling, Hans-Jürgen (2016): Die politische Theorie des Neo-Marxismus: Antonio Gramsci. In André Brodocz, Gary S. Schaal (Eds.): Politische Theorien der Gegenwart. Eine Einführung. 4., überarbeitete und aktualisierte Auflage. Opladen & Toronto: Verlag Barbara Budrich (UTB Politikwissenschaft, 2218), pp. 407–446.

Bilchit, David (2016): The Necessity for a Business and Human Rights Treaty. In *Bus. and hum. rights j.* 1 (2), pp. 203–227. DOI: https://doi.org/10.1017/bhj.2016.13.

Bitzer, Verena; Glasbergen, Pieter; LEROY, PIETER (2012): Partnerships of a feather flock together? An analysis of the emergence of networks of partnerships in the global cocoa sector. In *Global Networks* 12 (3), pp. 355–374. DOI: https://doi.org/10.1111/j.1471-0374.2011.00359.x.

Blowfield, Michael (2007): Reasons to Be Cheerful? What We Know about CSR's Impact. In *Third World Quarterly* 28 (4), pp. 683–695. Available online at http://www.jstor.org/stable/20454956.

Blowfield, Michael; Frynas, Jedrzej George (2005): Setting New Agendas: Critical Perspectives on Corporate Social Responsibility in the Developing World. In *International Affairs (Royal Institute of International Affairs 1944-)* 81 (3), pp. 499–513. Available online at http://www.jstor.org/stable/3569630.

Blowfield, Michael; Murray, Alan (2019): Corporate responsibility. Fourth edition. Oxford: Oxford University Press.

Blowfield, Michael E. (2003): Ethical Supply Chains in the Cocoa, Coffee and Tea Industries. In *Greener Management International* 3, pp. 14–24. Available online at http://www.iscom.nl/publicaties/gmi43blo[1].pdf.

Blowfield, Michael E.; Dolan, Catherine S. (2008): Stewards of Virtue? The Ethical Dilemma of CSR in African Agriculture. In *Development & Change* 39 (1), pp. 1–23. DOI: https://doi.org/10.1111/j.1467-7660.2008.00465.x.

Boas, Morten; Huser, Anne (2006): Child Labour and Cocoa Production in West Africa. The case of Côte d'Ivoire and Ghana. Edited by Institute for Applied International Studies,

Norway (Fafo). Available online at https://fafo.no/index.php/zoo-publikasjoner/fafo-rapporter/item/child-labour-and-cocoa-production-in-west-africa?tmpl=component&pri nt=1, checked on 9/7/2020.

Börzel, Tanja A.; Risse, Thomas (2010): Governance without a state: Can it work? In *Regulation & Governance* 4 (2), pp. 113–134. DOI: https://doi.org/10.1111/j.1748-5991.2010. 01076.x.

Brako, Dompreh Eric; Richard, Asare; Alexandros, Gasparatos (2020): Do voluntary certification standards improve yields and wellbeing? Evidence from oil palm and cocoa smallholders in Ghana. In *International Journal of Agricultural Sustainability*, pp. 1–24. DOI: https://doi.org/10.1080/14735903.2020.1807893.

Brand, Ulrich (2007): Die Internationalisierung des Staates als Rekonstruktion von Hegemonie. Zur staatstheoretischen Erweiterung Gramscis. In Sonja Buckel, Andreas Fischer-Lescano (Eds.): Hegemonie gepanzert mit Zwang. Zivilgesellschaft und Politik im Staatsverständnis Antonio Gramscis. 1. Auflage. Baden-Baden: Nomos Verlagsgesellschaft mbH & Co. KG, pp. 161–180.

Bregman, Rutger (2017): Utopia for Realists. And How We Can Get There. Paperback edition. London: Bloomsbury Publishing.

Buckel, Sonja; Fischer-Lescano, Andreas (Eds.) (2007): Hegemonie gepanzert mit Zwang. Zivilgesellschaft und Politik im Staatsverständnis Antonio Gramscis. 1. Auflage. Baden-Baden: Nomos Verlagsgesellschaft mbH & Co. KG.

Bundesministerium für wirtschaftliche Zusammenarbeit und Entwicklung (n.d.): Faire globale Liefer- und Wertschöpfungsketten. Available online at https://www.bmz.de/de/the men/lieferketten/index.html, checked on 10/4/2020.

Bundesministerium für wirtschaftliche Zusammenarbeit und Entwicklung (2019): Initiative für nachhaltige Agrarlieferketten (INA). Eine Plattform von Akteuren aus Privatwirtschaft, Zivilgesellschaft und Politik. Available online at https://www.bmz.de/de/ themen/lieferketten/index.html, checked on 10/4/2020.

Burnham, Peter (1991): Neo-Gramscian Hegemony and the International Order. In *Capital & Class* 15 (3), pp. 73–92. DOI: https://doi.org/10.1177/030981689104500105.

Candeias, Mario (2008): Gramscianische Konstellationen. Hegemonie und die Durchsetzung neuer Produktions- und Lebensweisen. In Andreas Merkens, Victor Rego Diaz (Eds.): Mit Gramsci arbeiten. Texte zur politisch-praktischen Aneignung Antonio Gramscis. Dt. Orig.-Ausg, 2. Auflage. Hamburg: Argument-Verl. (Argument-Sonderband. N.F, 305), pp. 15–32.

Cappelle, Jan (2008): Towards a Sustainable Cocoa Chain. Power and possibilities within the cocoa and chocolate sector. Edited by Oxfam International.

Carroll, Archie B. (1979): A Three-Dimensional Conceptual Model of Corporate Performance. In *The Academy of Management Review* 4 (4), p. 497. DOI: https://doi.org/10. 2307/257850.

Carroll, Archie B. (1999): Corporate Social Responsibility. In *Business & Society* 38 (3), pp. 268–295. DOI: https://doi.org/10.1177/000765039903800303.

Carroll, Archie B.; Shabana, Kareem M. (2010): The Business Case for Corporate Social Responsibility: A Review of Concepts, Research and Practice. In *Int J Management Reviews* 12 (1), pp. 85–105. DOI: https://doi.org/10.1111/j.1468-2370.2009.00275.x.

Central Intelligence Agency (n.d.): The World Fact Book. Ghana. Available online at https:// www.cia.gov/library/publications/the-world-factbook/geos/gh.html, checked on 10/1/

2020.

Charlery de la Masselière, Bernard (2002): Filières agricoles des produits tropicaux. In *Les Cahiers d'Outre-Mer* 55 (220), pp. 365–370. DOI: https://doi.org/10.4000/com.940.

Charlery de la Masselière, Bernard (2005): Le biais rural. Retour sur le développement. In *Outre-Terre* 11 (2), pp. 41–55. Available online at https://www.cairn.info/revue-outre-ter re1-2005-2-page-41.htm.

Charlery de la Masselière, Bernard (2014): Penser la question paysanne en Afrique intertropicale. Toulouse: Presse universitaire du Mirail (Ruralités nord-sud).

Cheyns, Emmanuelle; Riisgaard, Lone (2014): Introduction to the symposium. In *AGR HUM VALUES* 31 (3), pp. 409–423. DOI: https://doi.org/10.1007/s10460-014-9508-4.

Chocolate Manufacturer Association (2001): Harkin Engel Protocol. Available online at https://cocoainitiative.org/knowledge-centre-post/harkin-engel-protocol/, checked on 9/7/2020.

Clapp, Jennifer; Fuchs, Doris A. (Eds.) (2009): Corporate power in global agrifood governance. Cambridge, Mass.: MIT Press (Food, health, and the environment). Available online at http://search.ebscohost.com/login.aspx?direct=true&scope=site&db=nlebk& db=nlabk&AN=259236.

Clarence-Smith, William (2016): Chocolate Consumption from the Sixteenth Century to the Greag Chocolate Boom. In Mara P. Squicciarini, Johan F. M. Swinnen (Eds.): The economics of chocolate. Oxford: Oxford University Press, pp. 43–70.

Codjoe, Francis Nana Yaw; Ocansey, Charles K.; Boateng, Dennis O.; Ofori, Johnson (2013): Climate Change Awareness and Coping Strategies of Cocoa Farmers in Rural Ghana. In *Journal of Biology, Agriculture and Healthcare* 3 (11), pp. 19–29.

Coffee Circle (2020): Fair Trade Siegel im Vergleich. Available online at https://www.coffee circle.com/de/e/fair-trade-kritik, checked on 10/1/2020.

Commission of the European Communities (2001): Green Paper. Promoting a European framework for Corporate Social Responsibility. Brussels. Available online at https://www.eea.europa.eu/policy-documents/com-2001-366-final-green, checked on 8/30/2020.

Committee for Economic Development (1971): Social Responsibilities of Business Corporations. Available online at https://www.ced.org/reports/social-responsibilities-of-business-corporations, updated on 9/26/2020.

Cornwall, Andrea; Brock, Karen (2005): What do buzzwords do for development policy? a critical look at 'participation', 'empowerment' and 'poverty reduction'. In *Third World Quarterly* 26 (7), pp. 1043–1060. DOI: https://doi.org/10.1080/01436590500235603.

Cosbey, Aaron (2015): Policy case study – Food labelling Climate for Sustainable Growth. CEPS Carbon Market Forum, Centre for European Policy Studies. Brussels (Final Draft For Discussion Paper). Available online at http://papers.ssrn.com/sol3/papers.cfm?abs tract_id=2696716, checked on 5/6/2018.

Cox, Robert W. (1983): Gramsci, Hegemony and International Relations : An Essay in Method. In *Millennium* 12 (2), pp. 162–175. DOI: https://doi.org/10.1177/030582988301 20020701.

Crane, Andrew; Matten, Dirk; Moon, Jeremy (2009): Corporations and citizenship. Cambridge: Cambridge University Press.

Creswell, J. W. (1994): Research Design. Qualitative and Quantitative Approaches. Thousand Oaks, C.A.: SAGE Publications.

Curbach, Janina (2009): Die Corporate-Social-Responsibility-Bewegung. Zugl.: Bamberg, Univ., Diss., 2008. 1. Aufl. Wiesbaden: VS Verlag für Sozialwissenschaften (Wirtschaft + Gesellschaft). Available online at http://gbv.eblib.com/patron/FullRecord.aspx?p=748130.

Dei Antwi, Kwaku; Lyford, Richard, Conrad Power; Nartey, Yeboah (2018): Analysis of Food Security among Cocoa Producing Households in Ghana. In *Journal of Agriculture and Sustainability*.

Demirovic, Alex (2008): Neoliberalismus und Hegemonie. In Christoph Butterwegge (Ed.): Neoliberalismus. Analysen und Alternativen. 1. Aufl. Wiesbaden: VS Verl. für Sozialwiss, pp. 18–33.

Dengerink, Just (2013): Improving livelihoods with private sustainability standards: measuring the development impact of the UTZ Certified certification scheme among Ghanaian cocoa farmers. Masters thesis. Utrecht University. International Development.

Deppeler, Angela; Fromm, I.; Aidoo, Robert (2014): The Unmaking of the Cocoa Farmer: Analysis of Benefits and Challenges of third-party audited Certification Schemes for Cocoa Producers and Laborers in Ghana. International Conference on Food and Agribusiness Marketing Association. Cape Town, South Africa, 2014. Available online at https://www.bfh.ch/fileadmin/data/publikationen/2014/3_Deppeler_The_Unmaking_of_the_cocoal_farmer_IFAMA.pdf, checked on 5/6/2016.

Dicken, Peter (2015): Global shift. Mapping the changing contours of the world economy. 7. ed. New York, London: Guilford Press.

Dobner, Petra (2016): Wasserpolitik. Zur politischen Theorie, Praxis und Kritik globaler Governance. 2. Auflage. Berlin: Suhrkamp (Suhrkamp Taschenbuch Wissenschaft, 1958).

Dohmen, Manon Mireille; Helberg, Ulrich; Asiedu, Frank (2014): Certification Capacity Enhancement. Sustainable Cocoa Trainers' Manual. For Access to Certification and Increased Productivity. Ghana Version 2.0. Edited by German Initiative on Sustainable Cocoa. Available online at https://www.kakaoforum.de/en/our-work/certification-capacity-enhancement/, checked on 9/12/2020.

Dormon, E.N.A.; van Huis, A.; Leeuwis, C.; Obeng-Ofori, D.; Sakyi-Dawson, O. (2004): Causes of low productivity of cocoa in Ghana. Farmers' perspectives and insights from research and the socio-political establishment. In *NJAS – Wageningen Journal of Life Sciences* 52 (3–4), pp. 237–259. DOI: https://doi.org/10.1016/S1573-5214(04)80016-2.

Durry, Andrea; Schiffer, Thomas (2012): Kakao. Speise der Götter. München: Oekom-Verl. (Stoffgeschichten, Bd. 7).

Eberlein, Burkard (2019): Who Fills the Global Governance Gap? Rethinking the Roles of Business and Government in Global Governance. In *Organization Studies* 40 (8), pp. 1125–1145. DOI: https://doi.org/10.1177/0170840619847720.

Eberlein, Burkard; Abbott, Kenneth W.; Black, Julia; Meidinger, Errol; Wood, Stepan (2014): Transnational business governance interactions: Conceptualization and framework for analysis. In *Regulation & Governance* 8 (1), pp. 1–21. DOI: https://doi.org/10.1111/rego.12030.

Ellis, Frank (2000): The Determinants of Rural Livelihood Diversification in Developing Countries. In *Journal of Agricultural Economics* 51 (2), pp. 289–302. DOI: https://doi.org/10.1111/j.1477-9552.2000.tb01229.x.

Ellis, Karen; Keane, Jodie (2008): A review of ethical standards and labels: Is there a gap in the market for a new 'Good forDevelopment' label? Edited by Overseas Development Institute. London (ODI Working Paper, 297). Available online at https://www.odi. org/sites/odi.org.uk/files/odi-assets/publications-opinion-files/3332.pdf, checked on 5/6/ 2016.

European Center for Constitutional and Human Rights e.V. (2020): Hard law/soft law. Available online at https://www.ecchr.eu/glossar/hard-law-soft-law/.

Falk, Rainer (2020): Informationsbrief Weltwirtschaft & Entwicklung. Luxemburg (3– 5). Available online at https://www.weltwirtschaft-und-entwicklung.org/wearchiv/042 ae6ab520f6c006/042ae6ac0a095c001.php, checked on 10/4/2020.

Federal Ministry of Labour and Social Affairs (n.d.): CSR international. The EU's CSR policy. The Federal Government of Germany. Available online at https://www.csr-in-deutschland.de/EN/Policies/CSR-international/The-EUs-CSR-Policy/the-eus-csr-policy. html, checked on 9/26/2020.

Fenger, Nina Astrid; Bosselmann, Skovmand Aske; Asare, Richard; Neergaard, Andreas de (2017): The impact of certification on the natural and financial capitals of Ghanaian cocoa farmers. In *Agroecology and Sustainable Food Systems* 41 (2), pp. 143–166. DOI: https:// doi.org/10.1080/21683565.2016.1258606.

Fernandez-Stark, Karina; Gereffi, Gary (2011): Global Value Chains Analysis. A Primer. Duke University. Durham, North Carolina, USA.

Fischer, Karin (2020): Dependenz trifft Warenketten: Zur Überausbeutung von Arbeit im Globalen Süden. In *PROKLA. Zeitschrift für kritische Sozialwissenschaft* 50 (198), pp. 33–51.

Fischer, Karin; Reiner, Christian; Staritz, Cornelia (Eds.) (2010): Globale Güterketten. Weltweite Arbeitsteilung und ungleiche Entwicklung. Wien: Promedia; Südwind (Historische Sozialkunde : […], Internationale Entwicklung, 29).

Flick, Uwe; Kardorff, Ernst von; Steinke, Ines (2019): Was ist qualitative Forschung? Einleitung und Überblick. In Uwe Flick, Ernst von Kardorff, Ines Steinke (Eds.): Qualitative Forschung. Ein Handbuch. 13. Auflage, Originalausgabe. Reinbek bei Hamburg: Rowohlt Taschenbuch Verlag (Rororo, Rowohlts Enzyklopädie), pp. 13–29.

Fold, Niels (2001): Restructuring of the European chocolate industry and its impact on cocoa production in West Africa. In *Journal of Economic Geography* 1 (4), pp. 405–420. DOI: https://doi.org/10.1093/jeg/1.4.405.

Fold, Niels (2002): Lead Firms and Competition in 'Bi-polar' Commodity Chains: Grinders and Branders in the Global Cocoa-chocolate Industry. In *Journal of Agrarian Change* 2 (2), pp. 228–247.

Fold, Niels (2008): Transnational Sourcing Practices in Ghana's Perennial Crop Sectors. In *Journal of Agrarian Change* 8 (1), pp. 94–122.

Fold, Niels; Larsen, Marianne Nylandsted (2008): Globalization and restructuring of African commodity flows. Uppsala: Nordiska Afrikainstitutet.

Fold, Niels; Neilson, Jeff (2016): Sustaining Supplies in Smallholder-Dominated Value Chains. Corporate Governance of the Global Cocoa Sector. In Mara P. Squicciarini, Johan F. M. Swinnen (Eds.): The economics of chocolate. Oxford: Oxford University Press, pp. 195–212.

Fountain, Antonie; Hütz-Adams, Friedel (2015): Kakao-Barometer. Available online at http://www.cocoabarometer.org, checked on 9/26/2016.

Fountain, Antonie; Hütz-Adams, Friedel (2018): Cocoa Barometer. Available online at www. cocoabarometer.orf, checked on 8/29/2020.

Franzen, Margaret; Borgerhoff Mulder, Monique (2007): Ecological, economic and social perspectives on cocoa production worldwide. In *Biodivers Conserv* 16 (13), pp. 3835–3849. DOI: https://doi.org/10.1007/s10531-007-9183-5.

Frederick, William C. (1986): Toward CSR 3 : Why Ethical Analysis is Indispensable and Unavoidable in Corporate Affairs. In *California Management Review* 28 (2), pp. 126–141. DOI: https://doi.org/10.2307/41165190.

Frederick, William C. (1994): From CSR1 to CSR2. In *Business & Society* 33 (2), pp. 150–164. DOI: https://doi.org/10.1177/000765039403300202.

Frederick, William C. (1998): Moving to CSR. In *Business & Society* 37 (1), pp. 40–59. DOI: https://doi.org/10.1177/000765039803700103.

Fromm, Ingrid (2016): From Small Chocolatiers to Multinationals to Sustainable Sourcing. A Historical Review of the Swiss Chocolate Industry. In Mara P. Squicciarini, Johan F. M. Swinnen (Eds.): The economics of chocolate. Oxford: Oxford University Press, pp. 71–87.

Frynas, Jedrzej George (2005): The false developmental promise of Corporate Social Responsibility: evidence from multinational oil companies. In *International Affairs* 81 (3), pp. 581–598. DOI: https://doi.org/10.1111/j.1468-2346.2005.00470.x.

Frynas, Jędrzej George; Stephens, Siân (2015): Political Corporate Social Responsibility: Reviewing Theories and Setting New Agendas. In *International Journal of Management Reviews* 17 (4). Available online at http://oro.open.ac.uk/59667/.

Fuchs, Doris; Kalfagianni, Agni; Arentsen, Maarten (2009): Retail Power, Private Standards, and Sustainability. In Jennifer Clapp, Doris A. Fuchs (Eds.): Corporate power in global agrifood governance. Cambridge, Mass.: MIT Press (Food, health, and the environment), pp. 29–60.

Fuchs-Heinritz, Werner (2013): Partizipation. In Werner Fuchs-Heinritz, Daniela Klimke, Rüdiger Lautmann, Otthein Rammstedt, Urs Stäheli, Christoph Weischer, Hanns Wienold (Eds.): Lexikon zur Soziologie. 5., überarb. Aufl. 2011. Wiesbaden: Springer VS.

Gastinger, Karin; Gaggl, Philipp (2012): CSR als strategischer Managementansatz. In Andreas Schneider, René Schmidpeter (Eds.): Corporate Social Responsibility. Verantwortungsvolle Unternehmensführung in Theorie und Praxis. 2012nd ed. Berlin, Heidelberg: Springer Berlin Heidelberg, pp. 243–258.

Gayi, Samuel; Tsowou, Komi (2016): Cocoa Industry: Integrating small farmers into the global value chain. Edited by United Nations. UNCTAD Special Unit on Commodities. New York, Geneva (UNCTAD/SUC/2015/4). Available online at http://unctad.org/en/Pub licationsLibrary/suc2015d4_en.pdf, checked on 8/4/2017.

Genier, Claudia; Stamp, Mike; Pfitzer, Marc: Corporate Social Responsibility in the Agrifood Sector: Harnessing Innovation for Sustainable Development. Available online at https://de.scribd.com/document/200952185/CSR-in-the-Agrifood-Sector.

George Afrane; Rickard Arvidsson; Henrikke Baumann; Josefin Borg; Emma Keller; Llorenc Mila i Canals; Julie K Selmer (2013): A product chain organisation study of certified cocoa supply. In *6th International Conference on Life Cycle Management, LCM2013, 25–28 August 2013,Göteborg.* Available online at https://research.chalmers.se/en/public ation/177467.

Gereffi, Gary (1995): Global Production Systems and Third World Development. In Barbara Stallings (Ed.): Global change, regional response. The new international context of development. Cambridge, New York: Cambridge University Press, pp. 100–142.

Gereffi, Gary (1996a): Global commodity chains. New forms of coordination and control among nations and firms in international industries. In *Competition and Change* 1 (4), pp. 427–439.

Gereffi, Gary (1996b): Global commodity chains. New forms of coordination and control among nations and firms in international industries. In *Competition and Change* 1 (4), pp. 427–439.

Gereffi, Gary (2001): Beyond the Producerdriven/Buyer-driven DichotomyThe Evolution of Global Value Chains in the Internet Era (IDS Bulletin).

Gereffi, Gary (2014): Global value chains in a post-Washington Consensus world. In *Review of International Political Economy* 21 (1), pp. 9–37. DOI: https://doi.org/10.1080/096 92290.2012.756414.

Gereffi, Gary; Humphrey, John; Kaplinsky, Raphael; Sturgeon, Timothy J. (2001a): Introduction: Globalisation, Value Chains and Development. Institute of Development Studies (IDS Bulletin, 32.3).

Gereffi, Gary; Humphrey, John; Kaplinsky, Raphael; Sturgeon*, Timothy J. (2001b): Introduction: Globalisation, Value Chains and Development. In *1759–5436* 32 (3), pp. 1–8. DOI: https://doi.org/10.1111/j.1759-5436.2001.mp32003001.x.

Gereffi, Gary; Humphrey, John; Sturgeon, Timothy (2005): The governance of global value chains. In *Review of International Political Economy* 12 (1), pp. 78–104. DOI: https://doi. org/10.1080/09692290500049805.

Gereffi, Gary; Korzeniewicz, Miguel (Eds.) (1994): Commodity chains and global capitalism. ebrary, Inc. Westport, Connecticut: Praeger (Contributions in Economics & Economic History, 149, v.No. 149). Available online at http://site.ebrary.com/lib/alltitles/doc Detail.action?docID=10377260.

Germain, Randall (1998): Engaging Gramsci: international relations theory and the new Gramscians. In *Rev. Int. Stud.* 24 (1), pp. 3–21. DOI: https://doi.org/10.1017/S02602105 98000035.

Ghana Cocoa Board (n.d.): Regional Cocoa Purchases. Available online at https://cocobod. gh/cocoa-purchases, checked on 10/4/2020.

Ghana Cocoa Board (2013): Presentation on Ghana's National Cocoa Plan. Presented at: 88th Session, Wembley. Available online at https://www.icco.org/about-us/international-cocoa-agreements/cat_view/51-international-cocoa-council-and-subsidiary-bodies/70-international-cocoa-council-and-subsidiary-bodies-88th-session-wembley-london-23-27-september-2013.html, checked on 9/9/2020.

Ghana Cocoa Board (2014a): 45nd Annual Report & Financial Statements. Edited by Ghana Cocoa Board.

Ghana Cocoa Board (2014b): Implementation of the Global Cocoa Agenda: Progress report on Ghana's National Cocoa Plan. Available online at https://www.icco.org/about-us/international-cocoa-agreements/cat_view/81-world-cocoa-conference-amsterdam-2014/167-panel-1-tuesday-10-june.html, checked on 9/9/2020.

Ghana Cocoa Board (2016): 47th Annual Report and Financial Statement. Accra. Available online at https://cocobod.gh/resources/annual-report, updated on 2016, checked on 9/9/2020.

Ghana Cocoa Board (2020a): Ghana Cocoa Platform. Accra. Available online at http://www. ghanacocoaplatform.org/, checked on 9/9/2020.

Ghana Cocoa Board (2020b): Marketing and Statistics. Accra. Available online at https://coc obod.gh/resources/marketing-statistics, checked on 10/1/2020.

Ghana Cocoa Board (2020c): Subsidiaries and divisions. Accra. Available online at https:// cocobod.gh/subsidiaries-and-divisions, checked on 10/1/2020.

Ghana Cocoa Board (2020d): The main objectives of the board. Accra. Available online at https://cocobod.gh/objectives-of-board, checked on 10/1/2020.

Ghana Cocoa, Coffee, and Shea-Nut Farmers Association (n.d.): Welcome to Ghana Cocoa, Coffee, and Shea-Nut Farmers Association. Available online at http://www.whitehillinf ormationsystems.co.uk/CoffeeCocoa/welcome.html, checked on 10/1/2020.

Ghana Statistical Service (2012): 2010 Population and Housing Census. Summary Report of Final Results. Government of Ghana. Accra. Available online at https://statsghana. gov.gh/, checked on 9/22/2020.

Ghana Statistical Service (2016): Regional Spatial Business Report. Integrated Business Establishment Survey. Government of Ghana. Available online at http://statsghana.gov. gh/gssmain/fileUpload/pressrelease/REGIONAL%20SPATIAL%20BUSINESS%20R EPORT.pdf, checked on 9/22/2020.

Ghana Statistical Service (2018): Ghana Living Standards Survey Round 7. Poverty Trends in Ghana 2005–2017. Edited by Government of Ghana. Available online at https://www2. statsghana.gov.gh/nada/index.php/catalog/97, checked on 9/22/2020.

Gibbon, Peter; Bair, Jennifer; Ponte, Stefano (2008): Governing global value chains. An introduction. In Economy & Soc. 37 (3), pp. 315–338. DOI: https://doi.org/10.1080/030 85140802172656.

Gibbon, Peter; Ponte, Stefano (2008): Global value chains. From governance to governmentality? In Economy & Soc. 37 (3), pp. 365–392. DOI: https://doi.org/10.1080/030851408 02172680.

Gibbons, Steve (2011): Protect, Respect and Remedy: what does Ruggie's framework mean for ethical trade? Ethical Trading Initiative. Available online at https://www.ethicaltrade. org/blog/protect-respect-and-remedy-what-does-ruggies-framework-mean-ethical-trade, checked on 9/26/2020.

Gill, S. (2008): Power and Resistance in the New World Order. New York: Palgrave Macmillan.

Gill, Stephen (2011): Progressives politisches Handeln und die globale organische Krise. In Benjamin Opratko, Oliver Prausmüller (Eds.): Gramsci global. Neogramscianische Perspektiven in der internationalen politischen Ökonomie. Dt. Orig.-Ausg. Hamburg: Argument-Verl. (Argument Sonderband, N.F., 310), pp. 265–283.

Giovannucci, Daniele; Ponte, Stefano (2005): Standards as a new form of social contract? Sustainability initiatives in the coffee industry. In Food Policy 30 (3), pp. 284–301. DOI: https://doi.org/10.1016/j.foodpol.2005.05.007.

Glaser, Barney G.; Strauss, Anselm L. (1967): The discovery of grounded theory. Strategies for qualitative research. 4. paperback printing. New Brunswick [N.J.], London: Aldine-Transaction.

Gläser, Jochen; Laudel, Grit (2012): Experteninterviews und qualitative Inhaltsanalyse als Instrumente rekonstruierender Untersuchungen. 4. Auflage. Wiesbaden: VS, Verl. für Sozialwiss (Lehrbuch). Available online at http://d-nb.info/1002141753/04.

Glin, Laurent C.; Oosterveer, Peter; Mol, Arthur P.J. (2015): Governing the Organic Cocoa Network from Ghana. Towards Hybrid Governance Arrangements? In *Journal of Agrarian Change* 15 (1), pp. 43–64. DOI: https://doi.org/10.1111/joac.12059.

Global Interparliamentary Network (n.d.): Call of People's Representatives Worldwide. Representatives worldwide supporting the UN Binding Treaty on Transnational Corporations with respect to Human Rights. Binding Treaty.org. Available online at https://bindingtr eaty.org/#call, checked on 10/7/2020.

GlobeNewswire (n.d.): Global Chocolate Market Expected to Reach USD 161.56 Billion By 2024: Zion Market Research. New York. Available online at https://www.globenews wire.com/news-release/2018/10/22/1624439/0/en/Global-Chocolate-Market-Expected-to-Reach-USD-161-56-Billion-By-2024-Zion-Market-Research.html, checked on 10/1/2020.

Gockowski, Jim; Sonwa, Denis (2011): Cocoa intensification scenarios and their predicted impact on CO_2 emissions, biodiversity conservation, and rural livelihoods in the Guinea rain forest of West Africa. In *Environmental management* 48 (2), pp. 307–321. DOI: https://doi.org/10.1007/s00267-010-9602-3.

Grabosch, Robert; Scheper, Christian (2015): Corporate obligations with regard to human rights due diligence. Policy and legal approaches. Berlin: Friedrich-Ebert-Stiftung, Global Policy and Development (Study / Friedrich-Ebert-Stiftung).

Gramsci, Antonio (2012): Gefängnishefte. 1st ed. Edited by Klaus Bochmann, Wolfgang Fritz Haug. Hamburg: Argument Verl. (Argument / InkriT).

Hainmueller, Jens; Hiscox, Michael J.; Tampe, Maja (2011): Sustainable Development for Cocoa Farmers in Ghana. Baseline Survery: Preliminary report. MIT and Harvard University.

Hall, Steward (1991): Introductory Essay: Reading Gramsci. In Roger Simon (Ed.): Gramsci's political thought. An introduction. Completely rev. and reset. London: Lawrence & Wishart.

Hartmann, Martin; Offe, Claus (Eds.) (op. 2011): Politische Theorie und Politische Philosophie. Ein Handbuch. Originalausgabe. München: C. H. Beck (Beck'sche Reihe, 1819).

Hatanaka, Maki; Busch, Lawrence (2008): Third-Party Certification in the Global Agrifood System: An Objective or Socially Mediated Governance Mechanism? In *Sociologia Ruralis* 48 (1), pp. 73–91. DOI: https://doi.org/10.1111/j.1467-9523.2008.00453.x.

Hatanaka, Maki; Konefal, Jason (2013): Legitimacy and Standard Development in Multi-Stakeholder Initiatives. A Case Study of the Leonardo Academy's Sustainable Agriculture Standard Initiative. In *International Journal of Sociology of Agriculture and Food* 20 (2), pp. 155–173. Available online at http://www.ijsaf.org/archive/20/2/hatanaka.pdf, checked on 6/13/2016.

Henson, Spencer; Reardon, Thomas (2005): Private agri-food standards. Implications for food policy and the agri-food system. In *Food Policy* 30 (3), pp. 241–253. DOI: https://doi.org/10.1016/j.foodpol.2005.05.002.

Heydenreich, Cornelia; Guhr, Sarah (2020): Anforderungen an wirkungsvolle Multi-Stakeholder-Initiativen zur Stärkung unternehmerischer Sorgfaltspflichten. Empfehlungen aus Sicht der Zivilgesellschaft. CorA-Netzwerk für Unternehmensverantwortung. Berlin (Positionspapier). Available online at https://germanwatch.org/de/18894, checked on 10/4/2020.

Hill, Polly (1961): The Migrant Cocoa Farmers of Southern Ghana. In *Journal of the International African Institute* 31 (3), pp. 209–230.

Hopkins, Terence K.; Wallerstein, Immanuel (1977): Patterns of Development of the Modern World-System. In *Review (Fernand Braudel Center)* 1 (2), pp. 111–145. Available online at http://www.jstor.org/stable/40240765.

Horn, Laura (2011): "Aufgesogen" in den neoliberalen Konsensus? Trasformismo und Corporate Social Responsibility in der Europäischen Union. In Benjamin Opratko, Oliver Prausmüller (Eds.): Gramsci global. Neogramscianische Perspektiven in der internationalen politischen Ökonomie. Dt. Orig.-Ausg. Hamburg: Argument-Verl. (Argument Sonderband, N.F., 310), pp. 207–223.

Hulme, Mike (1992): Rainfall changes in Africa: 1931–1960 to 1961–1990. In *Int. J. Climatol.* 12 (7), pp. 685–699. DOI: https://doi.org/10.1002/joc.3370120703.

Human Rights Watch (2011): UN Human Rights Council: Weak Stance on Business Standards. Global Rules Needed, Not Just Guidance. Available online at https://www.hrw.org/news/2011/06/16/un-human-rights-council-weak-stance-business-standards, checked on 9/26/2020.

Humes, Edward (2011): Force of Nature. The Unlikely Story of Wal-mart's Green Revolution. New York: Harper Business.

Humphrey, John (2007): Upgrading in Global Value Chains.

Humphrey, John; Schmitz, Hubert (2001): Governance in Global Value Chains. Edited by Institute of Development Studies (IDS Bulletin, 32.3). Available online at https://www.ids.ac.uk/files/humphrey_schmitz_32_3.pdf, checked on 6/4/2016.

Hütz-Adams, Friedel (2009): Die dunklen Seiten der Schokolade: Große Preisschwankungen, schlechte Arbeitsbedingungen der Kleinbauern Langfassung. Edited by Südwind. Aachen. Available online at https://www.suedwind-institut.de/alle-verfuegbaren-publikationen/die_dunklen_seiten_der_schokolade_langfassung.html, checked on 9/7/2020.

Hütz-Adams, Friedel (2012): Vom Kakaobaum bis zum Konsumenten. Die Wertschöpfungskette von Schokolade. Siegburg: Südwind e.V.

Hütz-Adams, Friedel; Greiner, Clemens (2020): Kakaoanbau – Armut trotz Zertifizierung. Fairer Handel. In *Geographische Rundschau* 79 (1–2), pp. 38–42.

Iddrisu, Mubarak; Aidoo, Robert; Abawiera Wongnaa, Camillus (2020): Participation in UTZ-RA voluntary cocoa certification scheme and its impact on smallholder welfare: Evidence from Ghana. In *World Development Perspectives* 20, p. 100244. DOI: https://doi.org/10.1016/j.wdp.2020.100244.

Inglin, Andrea (2014): Mapping the Cocoa Cluster in Ghana. Edited by Swiss State Secretariat of Economic Affairs (SECO) (ETH Zürich, NADEL – Centre for Development and Cooperation).

Ingram, Verina; van Rijn, Fedes; Waarts, Yuca; Gilhuis, Henk (2018): The Impacts of Cocoa Sustainability Initiatives in West Africa. In *Sustainability* 10 (11), p. 4249. DOI: https://doi.org/10.3390/su10114249.

Initiative Lieferkettengesetz (n.d.): Anforderungen an ein wirksames Lieferkettengesetz. Berlin. Available online at https://lieferkettengesetz.de/forderungen/, checked on 10/4/2020.

INKOTA (2016): Die bittere Wahrheit über Schokolade. Available online at https://webshop.inkota.de/produkt/download-factsheet-inkota-infoblaetter/infoblatt-1-die-bittere-wahrheit-ueber-schokolade, checked on 9/7/2020.

INKOTA-netzwerk (2020a): Kakaoproduktion: ein Überblick. Make chocolate fair. Available online at http://de.makechocolatefair.org/themen/kakaoproduktion-ein-uberblick, checked on 9/26/2020.

INKOTA-netzwerk (2020b): Make chocolate fair. Demands. Available online at https://mak echocolatefair.org/demands, checked on 9/26/2020.

International Chamber of Commerce (2017): Response of the international business community to the "elements" for a draft egally binding instrument on transnational corporations and other business enterprises with respect to human rights. Available online at https://iccwbo.org/publication/un-treaty-business-human-rights-business-response-ele ments-draft-legally-binding-instrument/, checked on 10/7/2020.

International Cocoa Initiative (n.d.a): About the International Cocoa Initiative. Available online at https://cocoainitiative.org/about-ici/about-us/, checked on 10/1/2020.

International Cocoa Initiative (n.d.b): Our partners. Available online at https://cocoainitiat ive.org/about-ici/about-us/, checked on 10/1/2020.

International Cocoa Organization (2015): Quarterly Bulletin of Cocoa Statistics. Cocoa year 2014/15 (Vol. XLI, No. 3). Available online at https://www.icco.org/about-us/internati onal-cocoa-agreements/cat_view/30-related-documents/46-statistics-production.html, checked on 9/12/2020.

International Cocoa Organization (2017): Production of cocoa beans. Extract of ICCO Quarterly Bulletin of Cocoa Statistics, Vol. XLIII, No. 1, Cocoa year 2016/17. Available online at https://www.icco.org/statistics/production-and-grindings/production.html, checked on 4/26/2017.

International Cocoa Organization (2018): Berlin Declaration of the Fourth World Cocoa Conference. Abidjan. Available online at https://www.icco.org/about-us/icco-news/387-berlin-declaration-of-the-fourth-world-cocoa-conference.html, checked on 9/26/2020.

International Cocoa Organization (2020a): About ICCO. Abidjan. Available online at https://www.icco.org/about-us/about-the-icco.html, checked on 9/26/2020.

International Cocoa Organization (2020b): Council and Subsidiary Bodies. Abidjan. Available online at https://www.icco.org/about-us/council-and-subsidiary-bodies.html, checked on 9/26/2020.

International Cocoa Organization (2020c): The chocolate industry. Who are the main manufacturers of chocolate in the world? Abidjan. Available online at https://www.icco.org/about-cocoa/chocolate-industry.html, checked on 9/26/2020.

International Cocoa Organization (2020d): Quarterly Bulletin of Cocoa Statistics. Cocoa year 2019/20 (Vol. XLVI, No.2,). Available online at https://www.icco.org/about-us/int ernational-cocoa-agreements/cat_view/30-related-documents/46-statistics-production. html, checked on 9/12/2020.

International Institute for Sustainable Development (2020): About. Available online at https://www.iisd.org/ssi/about/, updated on 9/12/2020.

International Labour Organization (1998): ILO Declaration on Fundamental Principles and Rights at Work. Available online at https://www.ilo.org/declaration/lang--en/index.htm, checked on 9/26/2020.

International Law and Policy Institute (2015): Child Labour in the West African Cocoa Sector. Edited by International Law and Policy Institute.

International Monetary Fund (2009): Impact of the Global Financial Crisis on Sub-Saharan Africa (Miscellaneous Publication). Available online at https://www.imf.org/en/Public ations/Miscellaneous-Publication-Other/Issues/2016/12/31/Impact-of-the-Global-Financ ial-Crisis-on-Sub-Saharan-Africa-22745, checked on 9/9/2020.

International Organization for Standardization (2018): 26000 Guidance on Social Responsibility. Genève. Available online at https://www.iso.org/publication/PUB100258.html, checked on 8/30/2020.

Jaffee, Daniel (2012): Weak Coffee. Certification and Co-Optation in the Fair Trade Movement. In *Social Problems* 59 (1), pp. 94–116. DOI: https://doi.org/10.1525/sp.2012.59. 1.94.

Jones, Marc T. (2009): Disrobing the emperor: mainstream CSR research and corporate hegemony. In *Management of environmental quality* 20 (3), pp. 335–346. DOI: https://doi.org/ 10.1108/14777830910950720.

Jones, Steve (2006): Antonio Gramsci. London: Routledge (Routledge critical thinkers).

Kaplinsky, Raphael (2010): The Role Of Standards In Global Value Chains.

Kaplinsky, Raphael; Morris, Mike (2003): Handbook for Value Chain Research. International Development Research Center. Available online at http://www.ids.ac.uk/ids/global/ pdfs/VchNov01.pdf, checked on 4/30/2016.

Kilelu, Catherine; Klerkx, Laurens; Omore, Amos; Baltenweck, Isabelle; Leeuwis, Cees; Githinji, Julius (2017): Value Chain Upgrading and the Inclusion of Smallholders in Markets: Reflections on Contributions of Multi-Stakeholder Processes in Dairy Development in Tanzania. In *Eur J Dev Res* 29 (5), pp. 1102–1121. DOI: https://doi.org/10.1057/s41 287-016-0074-z.

Kloppers, Hendrik Jacobus (2012): Improving land reform through CSR : a legal framework analysis. With assistance of H. J. Pienaar: North-West University. Available online at https://repository.nwu.ac.za/handle/10394/8087.

Knierzinger, Johannes (2018): Bauxite mining in Africa. Transnational corporate governance and development. Cham, Switzerland: Palgrave Macmillan (International political economy series).

Knudsen, Michael Helt (2007): Making a living in the cocoa frontier, Western Ghana: Diversifying incomes in a cocoa economy. In *Geografisk Tidsskrift Danish Journal of Geography* 107 (2), pp. 29–44.

Knudsen, Michael Helt; Agergaard, Jytte (2015): Ghana's Cocoa Frontier in Transition. The role of migration and livelihoods diversification. In *Geografiska Annaler: Series B, Human Geography* 97 (4), pp. 325–342. DOI: https://doi.org/10.1111/geob.12084.

Knudsen, Michael Helt; Fold, Niels (2011): Land distribution and acquisition practices in Ghana's cocoa frontier. The impact of a state-regulated marketing system. In *Land Use Policy* 28 (2), pp. 378–387. DOI: https://doi.org/10.1016/j.landusepol.2010.07.004.

Kolavalli, Shashi; Vigneri, Marcella (2011): Cocoa in Ghana: Shaping the Success of an Economy. In Punam Chuhan-Pole, Manka Angwafo (Eds.): Yes Africa can. Success stories from a dynamic continent. Washington, DC: World Bank.

Kolavalli, Shashi; Vigneri, Marcella (2018): Cocoa Coast. The Board managed Cocoa Sector in Ghana. Washington, DC.

Kolavalli, Shashidhara; Vigneri, Marcella (2019): The Cocoa Coast: The Board-Managed Cocoa Sector in Ghana. In *Food Sec.* 11 (3), pp. 753–755. DOI: https://doi.org/10.1007/ s12571-019-00921-2.

Kolavalli, Shashidhara; Vigneri, Marcella; Maamah, Haruna; Poku, John (2012): The Partially Liberalized Cocoa Sector in Ghana. Producer Price Determination, Quality Control, and Service Provision. International Food Policy Research Institute. Accra, Ghana (IFPRI Discussion Paper, 01213).

KPMG (2012): The chocolate of tomorrow. What today's market can tell usabout the future. Edited by Haymarket Network Ltd.

KPMG (2014): A taste of the future. The trends that could transform the chocolate industry. Edited by Haymarket Network Ltd. Available online at https://home.kpmg.com/xx/en/home/insights/2014/06/taste-of-the-future.html, checked on 2/4/2014.

Kreide, Regina (2003): Poverty and Responsibility in a Globalized World. In *Analyse & Kritik* 25, pp. 199–219.

Kreide, Regina (2007): The Obligations of Transnational Corporations in the Global Context. In *Ethics and Economics* 4 (2), pp. 1–19.

Kreide, Regina (2009): Weltarmut und die Verpflichtungen kollektiver Akteure. In Barbara Bleisch (Ed.): Weltarmut und Ethik. 2., durchges. Aufl. Paderborn: Mentis (Ethica, 13), pp. 267–295.

Kreide, Regina (2019): Globale (Un)gerechtigkeiten? In Julian Nida-Rümelin, Detlef Daniels, Nicole Wloka (Eds.): Internationale Gerechtigkeit und institutionelle Verantwortung: De Gruyter, pp. 275–292.

Kuapa Kokoo (2017): About Kuapa Kokoo. Kumasi. Available online at https://www.kuapakokoo.com/index.php/about-us/, checked on 10/1/2020.

La Via Campesina (2007): Declaration of Nyéléni. Available online at https://www.nyeleni.org/spip.php?article290, checked on 10/4/2020.

Laclau, Ernesto; Mouffe, Chantal (2020): Hegemonie und radikale Demokratie. Zur Dekonstruktion des Marxismus. 6., überarbeitete Auflage. Edited by Michael Hintz, Gerd Vorwallner. Wien: Passagen Verlag (Passagen Philosophie).

Läderach, P.; Martinez-Valle, A.; Schroth, G.; Castro, N. (2013): Predicting the future climatic suitability for cocoa farming of the world's leading producer countries, Ghana and Côte d'Ivoire. In *Climatic Change* 119 (3–4), pp. 841–854. DOI: https://doi.org/10.1007/s10584-013-0774-8.

Laven, Anna (2007): Marketing reforms in Ghana's cocoa sector. Partial liberalisation, partial benefits? Edited by Overseas Development Institute (ODI Background Notes – Discussion papers).

Laven, Anna (2010): The risks of inclusion. Shifts in governance processes and upgrading opportunities for small-scale cocoa farmers in Ghana. Amsterdam: KIT.

Laven, Anna; Tyszler, Marcelo (2018): Demystifying the Cocoa Sector in Ghana and Côte d'Ivoire. Edited by KIT Royal Tropical Institute. Available online at https://www.kit.nl/project/demystifying-cocoa-sector/, checked on 9/7/2020.

Leal, Pablo Alejandro (2007): Participation: the ascendancy of a buzzword in the neo-liberal era. In *Development in Practice* 17 (4–5), pp. 539–548. DOI: https://doi.org/10.1080/09614520701469518.

Lee, Min-Dong Paul (2008): A review of the theories of corporate social responsibility: Its evolutionary path and the road ahead. In *Int J Management Reviews* 10 (1), pp. 53–73. DOI: https://doi.org/10.1111/j.1468-2370.2007.00226.x.

Leissle, Kristy (2018): Cocoa. Cambridge: Polity Press (Resources series).

Lemeilleur, Sylvaine; Ruf, François; N'Dao, Youssoupha (2015): The productivist rationality behind a sustainable certification process: Evidence from the Rainforest Alliance in the Ivorian cocoa sector. In *International Journal of Sustainable Development* 18 (4), pp. 310–328.

Lenssen, Gilbert; Blowfield, Michael E. (2005): Going global. How to identify and manage societal expectations in supply chains (and the consequences of failure). In *Corporate Governance* 5 (3), pp. 119–128. DOI: https://doi.org/10.1108/14720700510604751.

Lernoud, Julia; Potts, Jason; Sampson, Gregory; Schlatter, Bernhard; Huppe, Gabriel; Voora, Vivek et al. (2018): The State of Sustainable Markets 2018. Statistics and emerging trends. Edited by International Trade Centre (ITC), International Institute for Sustainable (IISD), Research Institute of Organic. Geneva. Available online at https://www.intracen.org/publication/The-State-of-Sustainable-Markets-2018-Statistics-and-Emerging-Trends/, checked on 9/12/2020.

Levy, David L. (2008): Political Contestation in Global Production Networks. In *The Academy of Management Review* 33 (4), pp. 943–963. DOI: https://doi.org/10.2307/20159454.

Levy, David L.; Newell, Peter J. (2002): Business Strategy and International Environmental Governance: Toward a Neo-Gramscian Synthesis. In *Global Environmental Politics* 2 (4), pp. 84–101. Available online at https://econpapers.repec.org/article/tprglenvp/v_3a2_3ay_3a2002_3ai_3a4_3ap_3a84-101.htm.

Levy, David L.; Szejnwald Brown, Halina; Jong, Martin de (2010): The Contested Politics of Corporate Governance. In *Business & Society* 49 (1), pp. 88–115. DOI: https://doi.org/10.1177/0007650309345420.

Loew, Thomas; Rohde, Friederike (2013): CSR und Nachhaltigkeitsmanagement Definitionen, Ansätze und organisatorische Umsetzung im Unternehmen. Institute for Sustainability. Berlin. Available online at https://www.esf-querschnittsziele.de/sucher gebnis detail/article//csr-und-nachhaltigkeitsmanagement-definitionen-ansaetze-und-org anisatorische-umsetzung-im-unternehmen.html, checked on 9/10/2020.

Long, John C. (2008): From Cocoa to CSR: Finding sustainability in a cup of hot chocolate. In *Thunderbird Int'l Bus Rev* 50 (5), pp. 315–320. DOI: https://doi.org/10.1002/tie.20215.

Lopez, Carlos (2019): The Revised Draft of a Treaty on Business and Human Rights: Ground-breaking improvements and brighter prospects. International Institute for Sustainable Development (IISD) (Investment Treaty News), updated on https://www.iisd.org/itn/10/2/2019/, checked on 10/7/2020.

Mäkinen, Jukka; Kasanen, Eero (2015): In defense of a regulated market economy. In *Journal of Global Ethics* 11 (1), pp. 99–109. DOI: https://doi.org/10.1080/17449626.2015.1004464.

Mäkinen, Jukka; Kasanen, Eero (2016): Boundaries Between Business and Politics: A Study on the Division of Moral Labor. In *Journal of Business Ethics* 134 (1), pp. 103–116. DOI: https://doi.org/10.1007/s10551-014-2419-x.

Mäkinen, Jukka; Kourula, Arno (2012): Pluralism in Political Corporate Social Responsibility. Available online at https://philpapers.org/rec/MKIPIP.

Marens, Richard (2007): Hollowing out of Corporate Social Responsibility: Abandoning a Tradition in an Age of Declining Hegemony, The. In *McGeorge Law Review* 39 (3). Available online at https://scholarlycommons.pacific.edu/mlr/vol39/iss3/11.

Matten, Dirk; Crane, Andrew (2005): Corporate Citizenship: Toward an Extended Theoretical Conceptualization. In *The Academy of Management Review* 30 (1), pp. 166–179. DOI: https://doi.org/10.2307/20159101.

Mayer, Frederick; Gereffi, Gary (2010): Regulation and Economic Globalization. Prospects and Limits of Private Governance. In *Business and Politics* 12 (3). DOI: https://doi.org/10.2202/1469-3569.1325.

Mayntz, Renate (2004): Governance im modernen Staat. In Arthur Benz (Ed.): Governance – Regieren in komplexen Regelsystemen. Eine Einführung. 1. Aufl. Wiesbaden: VS, Verl. für Sozialwiss (Governance, Bd. 1), pp. 71–83.

Mayring, Philipp (2015): Qualitative Inhaltsanalyse. Grundlagen und Techniken. 12., aktualisierte und überarb. Aufl. Weinheim: Beltz (Beltz Pädagogik). Available online at http://content-select.com/index.php?id=bib_view&ean=9783407293930.

McBarnet, Doreen (2009): Corporate Social Responsibility Beyond Law, Through Law, for Law: The New Corporate Accountability.

McMichael, Philip; Porter, Christine (2018): Going Public with Notes on Close Cousins, Food Sovereignty, and Dignity. In *1* 8 (A), pp. 207–212. DOI: https://doi.org/10.5304/jafscd.2018.08A.015.

Meinefeld, Werner (2019): Hypothesen und Vorwissen in der qualitativen Sozialforschung. In Uwe Flick, Ernst von Kardorff, Ines Steinke (Eds.): Qualitative Forschung. Ein Handbuch. 13. Auflage, Originalausgabe. Reinbek bei Hamburg: Rowohlt Taschenbuch Verlag (Rororo, Rowohlts Enzyklopädie), 265–175.

Melanie Coni-Zimmer (2012): Zivilgesellschaftliche Kritik und Corporate Social Responsibility als unternehmerische Legitimitätspolitik. In Anna Geis, Frank Nullmeier, Christopher Daase (Eds.): Der Aufstieg der Legitimitťspolitik. Rechtfertigung und Kritik politisch-konomischer Ordnungen. Der Aufstieg der Legitimitätspolitik. 1. Auflage 2012. Baden-Baden: Nomos Verlagsgesellschaft mbH & Co. KG, pp. 319–336, checked on 9/6/2012.

Mende, Janne (2017): Unternehmen als gesellschaftliche Akteure: Die unternehmerische Verantwortung für Menschenrechte zwischen privater und öffentlicher Sphäre. In *MenschenRechtsMagazin* 22 (1), pp. 5–17.

Mende, Janne (2020): Business Authority in Global Governance: Beyond Public and Private. WZB Berlin Social Science Center (Discussion Paper, No. SP IV).

Merkens, Andreas; Diaz, Victor Rego (Eds.) (2008): Mit Gramsci arbeiten. Texte zur politisch-praktischen Aneignung Antonio Gramscis. Dt. Orig.-Ausg, 2. Auflage. Hamburg: Argument-Verl. (Argument-Sonderband. N.F, 305).

Metzger, Jonas; Ollendorf, Franziska; Siebert, Anne (2016): Smallholder Farming under Threat? GIZ (Digital Development Debates, 16 Food and Farming). Available online at http://www.digital-development-debates.org/issue-16-food-farming--tradition--smallh older-farming-under-threat.html, checked on 10/12/2020.

Mikell, Gwendolyn (1989): Cocoa and chaos in Ghana. 1st ed. New York: Paragon House (A PWPA book).

Moir, Lance (2001): What do we mean by corporate social responsibility?: MCB UP Ltd (Emerald). Available online at https://dspace.lib.cranfield.ac.uk/handle/1826/3256.

Monastyrnaya, Elena; Joerin, Jonas; Dawoe, Evans; Six, Johann (2016): Assessing the resilience of the cocoa value chain in Ghana. Case study report. Edited by Swiss Federal Institute of Technology Zurich, ETH. Zürich.

Moon, Jeremy (2002): The Social Responsibility of Business and New Governance. In *Gov. & oppos.* 37 (3), pp. 385–408. DOI: https://doi.org/10.1111/1477-7053.00106.

Mooney, Pat (2018): Blocking the chain. Industrial Food Chain Concentration, big data platforms and food sovereignty solutions. ETC Group. Available online at https://www.etc group.org/content/blocking-chain, checked on 10/4/2020.

Morton, Adam David (2006): The Grimly Comic Riddle of Hegemony in IPE: Where is Class Struggle? In *Politics* 26 (1), pp. 62–72. DOI: https://doi.org/10.1111/j.1467-9256.2006. 00252.x.

Muilerman, Sander; Vellema, Sietze (2017): Scaling service delivery in a failed state: cocoa smallholders, Farmer Field Schools, persistent bureaucrats and institutional work in Côte d'Ivoire. In *International Journal of Agricultural Sustainability* 15 (1), pp. 83–98. DOI: https://doi.org/10.1080/14735903.2016.1246274.

Muojama, Olisa G. (2016): The First World War and Cocoa Industry in Ghana: A Study of the Hazards of Economic Dependency. In *Global Journal of Human-Social Science Research*, pp. 32–41.

Naden, Clare (2016): Big step forward for the cocoa sector with new global standards in the pipeline. International Organization for Standardization. Available online at https://www. iso.org/news/2016/09/Ref1936.html, checked on 10/1/2020.

Naden, Clare (2019): First International Standards for sustainable and traceable cocoa just published. International Organization for Standardization. Available online at https:// www.iso.org/news/2016/09/Ref1936.html, checked on 10/1/2020.

Nadvi, K. (2008): Global standards, global governance and the organization of global value chains. In *J Econ Geogr* 8 (3), pp. 323–343. DOI: https://doi.org/10.1093/jeg/lbn003.

Nadvi, Khalid; Wältring, Frank (2002): Making Sense of Global Standards. Gerhard Mercator-Universität Duisburg. Duisburg (INEF Report, 58).

Neilson, Jeff (2008): Global Private Regulation and Value-Chain Restructuring in Indonesian Smallholder Coffee Systems. In *World Development* 36 (9), pp. 1607–1622. DOI: https:// doi.org/10.1016/j.worlddev.2007.09.005.

Neilson, Jeffrey; Pritchard, Bill (2007): Green Coffee? The Contradictions of Global Sustainability Initiatives from an Indian Perspective. In *Development Policy Review* 25 (3), pp. 311–331. DOI: https://doi.org/10.1111/j.1467-7679.2007.00372.x.

Nelson, Valerie; Tallontire, Anne (2014): Battlefields of ideas. Changing narratives and power dynamics in private standards in global agricultural value chains. In *Agric Hum Values* 31 (3), pp. 481–497. DOI: https://doi.org/10.1007/s10460-014-9512-8.

Neuhäuser, Christian; Buddeberg, Eva (2015): Einleitung: Pflicht oder Verantwortung? In *Zeitschrift für Praktische Philosophie* 2 (2), pp. 49–60. Available online at https:// www.praktische-philosophie.org/uploads/8/0/5/2/80520134/zfpp.2.2015.einleitung.neu haeuser.buddenberg.pdf, checked on 8/17/2020.

Newell, Peter; Frynas, Jedrzej George (2007): Beyond csr ? Business, poverty and social justice: an introduction. In *Third World Quarterly* 28 (4), pp. 669–681. DOI: https://doi.org/ 10.1080/01436590701336507.

Noble, Mark D. (2017): Chocolate and The Consumption of Forests: A Cross-National Examination of Ecologically Unequal Exchange in Cocoa Exports. In *JWSR* 23 (2), pp. 236–268. DOI: https://doi.org/10.5195/jwsr.2017.731.

OECD (2012): OECD Guidelines for Multinational Enterprises. Edited by OECD Publishing. Available online at https://doi.org/10.1787/9789264115415-en, checked on 8/30/

2020.

Ofosu-Asare, Kwaku (2011): Trade Liberalisation, Globalisation and the Cocoa Industry in Ghana: the case of the smallholder cocoa farmers. PhD Thesis. University of Westminster, Westminster. School of Social Sciences, Humanities and Languages. Available online at https://westminsterresearch.westminster.ac.uk/item/8zz99/trade-liberalisation-globalisation-and-the-cocoa-industry-in-ghana-the-case-of-the-smallholder-cocoa-far mers, checked on 9/9/2020.

Okoye, Adaeze (2009): Theorising Corporate Social Responsibility as an Essentially Contested Concept: Is a Definition Necessary? In J Bus Ethics 89 (4), pp. 613–627. DOI: https://doi.org/10.1007/s10551-008-0021-9.

Olayiwola, O. O. (2013): Livelihood Diversification: a concept note on marginal farmers driving forces in Africa. In Abhinav, International Monthly Refereed Journal of Research In Management and Technology 2, pp. 28–34.

Ollendorf, Franziska (2017): Certification privée pour la production de cacao durable et ses transformations locales. 10 pages. In Systèmes alimentaires 2017 (n° 2), pp. 103–112. DOI: https://doi.org/10.15122/isbn.978-2-406-07196-9.p.0103.

Ollendorf, Franziska (forthcoming, 2021): Corporate Social Responsibility in the Global Cocoa Chocolate Chain– Insights from sustainability certification in Ghana's cocoa communities. In Andrea Komlosy, Goran Music (Eds.): Global Commodity Chains and Labour Relations. Leiden: Brill (Studies in Global Social History).

Onumah, Justina Adwoa; Al-Hassan, Ramatu Mahama; Onumah, Edward Ebo (2013): Productivity and Efficiency of Cocoa Production in Eastern Ghana. In Journal of Economics and Sustainable Development 4 (4), pp. 106–117.

Oomes, Nienke; Tieben, Bert; et al. (2016): Market Concentration and Price Formation in the Global Cocoa Value Chain. no. 2016–79. SEO Amsterdam Economics. Amsterdam (SEO-report). Available online at http://www.seo.nl/uploads/media/2016-79_Market_Concentration_and_Price_Formation_in_the_Global_Cocoa_Value_Chain.pdf, checked on 8/4/2017.

Open Ended Inter-governmental Working Group (2018): Legally Binding Instrument to Regulate, in International Human Rights Law, the activities of Transnational Corporations and other Business Enterprises. Zero draft. United Nations Human Rights Council. Available online at https://bindingtreaty.org/resources/, checked on 10/7/2020.

Open Ended Inter-governmental Working Group (2019): Legally Binding Instrument to Regulate, in International Human Rights Law, the activities of Transnational Corporations and other Business Enterprises. OEIGWG chairmenship revised draft. United Nations Human Rights Council. Available online at https://bindingtreaty.org/resources/, checked on 10/7/2020.

Oppong, Francis (2016): Ghana GovernmentPolicies towards accelerated Growth in Cocoa Production. Edited by International Cocoa Organization. Ghana Cocoa Board.

Opratko, Benjamin; Prausmüller, Oliver (2011): Neogramscianische Perspektiven in der IPÖ: Eine Einführung. In Benjamin Opratko, Oliver Prausmüller (Eds.): Gramsci global. Neogramscianische Perspektiven in der internationalen politischen Ökonomie. Dt. Orig.-Ausg. Hamburg: Argument-Verl. (Argument Sonderband, N.F., 310), pp. 11–38.

Ougaard, Morten; Leander, Anna (Eds.) (2012): Business and Global Governance. Hoboken: Taylor and Francis (Warwick Studies in Globalisation).

Overbeek, Henk (2003): Transnational historical materialism: theories of transnational class formation and world order. In Ronen Palan (Ed.): Global political economy. Contemporary theories. Repr. London: Routledge (Routledge/RIPE studies in global political economy), pp. 168–183.

Palpacuer, Florence (2008): Bringing the social context back in. Governance and wealth distribution in global commodity chains. In *Economy & Soc.* 37 (3), pp. 393–419. DOI: https://doi.org/10.1080/03085140802172698.

Pattberg, Philipp (2006): The Influence of Global Business Regulation. Beyond Good Corporate Conduct. In *Business & Society Review* 111 (3), pp. 241–268. DOI: https://doi.org/ 10.1111/j.1467-8594.2006.00271.x.

Pattberg, Philipp; Widerberg, Oscar (2016): Transnational multistakeholder partnerships for sustainable development: Conditions for success. In *Ambio* 45 (1), pp. 42–51. DOI: https://doi.org/10.1007/s13280-015-0684-2.

Poelmans, Eline; Swinnen, Johan (2016): A Brief Economic History of Chocolate. In Mara P. Squicciarini, Johan F. M. Swinnen (Eds.): The economics of chocolate. Oxford: Oxford University Press, pp. 11–42.

Pogge, Thomas (2003): Armenhilfe'ins Ausland. In *Analyse & Kritik* 25 (2), pp. 202–247. Available online at https://philpapers.org/rec/POGAA.

Pogge, Thomas (2005): Severe Poverty as a Violation of Negative Duties. In *Ethics int. aff.* 19 (1), pp. 55–83. DOI: https://doi.org/10.1111/j.1747-7093.2005.tb00490.x.

Pogge, Thomas (2009): Severe Poverty as a Human Rights Violation. In Thomas Winfried Menko Pogge (Ed.): Freedom from poverty as a human right. Who owes what to the very poor? Repr. Oxford, Paris, UNESCO: Oxford University Press (Philosopher's Library series), pp. 11–54.

Ponte, Stefano (2019): Business, power and sustainability in a world of global value chains. London, UK: Zed.

Ponte, Stefano; Gibbon, Peter (2005): Quality standards, conventions and the governance of global value chains. In *Economy and Society* 34 (1), pp. 1–31. DOI: https://doi.org/10. 1080/03085140420000329315.

Ponte, Stefano; Sturgeon, Timothy (2014): Explaining governance in global value chains. A modular theory-building effort. In *Review of International Political Economy* 21 (1), pp. 195–223. DOI: https://doi.org/10.1080/09692290.2013.809596.

Potts, Jason; Lynch, Matthew; Wilkings, Ann; Huppé, Gabriel A.; Cunningham, Maxine; Voora, Vivek Anand (2014): The state of sustainability initiatives review 2014. Standards and the green economy. Winnipeg, Manitoba, London [England], Beaconsfield, Quebec: International Institute for Sustainable Development; International Institute for Environment and Development; Canadian Electronic Library. Available online at https://www. iisd.org/pdf/2014/ssi_2014.pdf, checked on 4/30/2016.

Project Coordination Unit (2014): Detailed info on organisations visited. unpublished document. Ghana Cocoa Board. Accra.

Pye, Oliver (2017): A Plantation Precariat: Fragmentation and Organizing Potential in the Palm Oil Global Production Network. In *Development and Change* 48 (5), pp. 942–964. DOI: https://doi.org/10.1111/dech.12334.

Raikes, Philip; Friis Jensen, Michael; Ponte, Stefano (2000): Global commodity chain analysis and the French filière approach: comparison and critique. In *Economy & Soc.* 29 (3), pp. 390–417. DOI: https://doi.org/10.1080/03085140050084589.

Rainforest Alliance (2018): New Tools for Climate-Smart Cocoa Farming in Ghana, 5/ 31/2018. Available online at https://www.rainforest-alliance.org/article/new-tools-for-cli mate-smart-cocoa-farming-ghana, checked on 10/1/2020.

Rainforest Alliance (2019): The Rainforest Alliance Launches Cocoa Assurance Plan in West Africa. Available online at https://www.rainforest-alliance.org/articles/rainforest-all iance-launches-cocoa-assurance-plan-in-west-africa, checked on 10/11/2020.

Rainforest Alliance (2020a): 2020 Zertifizierungsprogramm. Available online at https:// www.rainforest-alliance.org/business/de/#why-merge, checked on 10/1/2020.

Rainforest Alliance (2020b): Cocoa Certification Data Report 2019. Rainforest Alliance and UTZ programs. Available online at https://www.rainforest-alliance.org/business/res ource-item/cocoa-certification-data-report-2019/, checked on 9/12/2020.

Ratjen, Sandra (2008): Food sovereignty and right to food: the case of Uganda. In *Rural 21. The International Journal for Rural Development*, pp. 25–27, checked on 10/4/2020.

Reardon, Thomas; Codron, Jean-Marie; Busch, Lawrence; Bingen, R. James; Harris, Craig (2001): Global Change in Agrifood Grades and Standards: Agribusiness strategic responses in developing countries. In *International Food and Agribusiness Management Review* 2 (1030–2016–83529). DOI: https://doi.org/10.22004/ag.econ.34227.

Rosenau, James N.; Czempiel, Ernst Otto (1992): Governance without government. Order and change in world politics. Cambridge, New York: Cambridge University Press (Cambridge studies in international relations, 20).

Ross, Corey (2014): The plantation paradigm: colonial agronomy, African farmers, and the global cocoa boom, 1870s–1940s. In *Journal of Global History* 9 (01), pp. 49–71. DOI: https://doi.org/10.1017/S1740022813000491.

Ruf, F. (1995): Booms et crises du cacao. Les vertiges de l'or brun. Montpellier: CIRAD-SAR/Paris: Karthala.

Ruf, Francois; N'Dao, Youssoupha; Lemeilleur, Sylvaine (2015a): Certification du cacao, stratégie à hauts risques. Available online at http://www.inter-reseaux.org/publications/ autres-publications/article/certification-of-cocoa-a-high-risk?lang=fr, checked on 10/4/ 2020.

Ruf, François (2007): The new Ghana cocoa boom in the 2000s : from forest clearing to Green revolution. CIRAD / University of Ghana.

Ruf, François; Bourgeois, Marion (2014): Pénurie de cacao ? Agro-industrie et planteurs villageois. Montpeillier. Available online at https://agritrop.cirad.fr/575375/1/document_ 575375.pdf.

Ruf, François; Kone, Siaka; Bebo, Boniface (2019a): Le boom de l'anacarde en Côte d'Ivoire : transition écologique et sociale des systèmes à base de coton et de cacao. In *Cah. Agric.* 28, p. 21. DOI: https://doi.org/10.1051/cagri/2019019.

Ruf, François; Leitz, Enrique Uribe; Gboko, Kouamé; Carimentrand, Aurélie (2019b): Des certifications inutiles ? Les relations asymétriques entre coopératives, labels et cacaocul-teurs en Côte d'Ivoire. In *Revue internationale des etudes du developpement* (4), pp. 31–61.

Ruf, François; Schroth, Götz; Doffangui, Kone (2015b): Climate change, cocoa migrations and deforestation in West Africa. What does the past tell us about the future? In *Sustain Sci* 10 (1), pp. 101–111. DOI: https://doi.org/10.1007/s11625-014-0282-4.

Ruf, Francois Olivier (2011): The Myth of Complex Cocoa Agroforests: The Case of Ghana. In *Human ecology: an interdisciplinary journal* 39 (3), pp. 373–388. DOI: https://doi.org/10.1007/s10745-011-9392-0.

Sabadoz, Cameron (2011): Between Profit-Seeking and Prosociality: Corporate Social Responsibility as Derridean Supplement. In *Journal of Business Ethics* 104 (1), pp. 77–91. DOI: https://doi.org/10.1007/s10551-011-0890-1.

Sanial, Elsa (2019): À la recherche de l'ombre, géographie des systèmes agroforestiers émergents en cacaoculture ivoirienne post-forestière. Dissertation. Université de Lyon, Lyon. Available online at https://www.researchgate.net/publication/338549035_A_la_r echerche_de_l'ombre_geographie_des_systemes_agroforestiers_emergents_en_cacaoc ulture_ivoirienne_post-forestiere, checked on 10/4/2020.

Sanial, Elsa; Ruf, François (2018): Is kola Tree the Enemy of Cocoa? A Critical Analysis of Agroforestry Recommendations Made to Ivorian Cocoa Farmers. In *Hum Ecol* 46 (2), pp. 159–170. DOI: https://doi.org/10.1007/s10745-018-9975-0.

Scherer, Andreas Georg (2018): Theory Assessment and Agenda Setting in Political CSR: A Critical Theory Perspective. In *Int J Management Reviews* 20 (2), pp. 387–410. DOI: https://doi.org/10.1111/ijmr.12137.

Scherer, Andreas Georg; Palazzo, Guido (2008): Globalization and Corporate Social Responsibility. In : The Oxford Handbook of Corporate Social Responsibility. Available online at https://www.oxfordhandbooks.com/view/https://doi.org/10.1093/oxfordhb/978019921 1593.001.0001/oxfordhb-9780199211593-e-018?rskey=4twhlo&result=1.

Scherer, Andreas Georg; Palazzo, Guido (2011): The New Political Role of Business in a Globalized World – A Review of a New Perspective on CSR and its Implications for the Firm, Governance, and Democracy.

Scherer, Andreas Georg; Rasche, Andreas; Palazzo, Guido; Spicer, André (2016): Managing for Political Corporate Social Responsibility: New Challenges and Directions for PCSR 2.0. In *Journal of Management Studies* 53 (3), pp. 273–298. DOI: https://doi.org/10.1111/joms.12203.

Scherrer, Christoph (1998): Neo-gramscianische Interpretationen internationaler Beziehungen. In Uwe Hirschfeld (Ed.): Gramsci-Perspektiven. Beiträge zur Gründungskonferenz des "Berliner Instituts für Kritische Theorie" vom 18. bis 20. April 1997 im Jagdscloss Glienicke, Berlin. Berlin: Argument Verlag (Argument Sonderband, n.F., AS 256), pp. 160–174.

Scherrer, Christoph (2005): Internationale Politische Ökonomie. In Wolfgang Fritz Haug, Frigga Haug, Peter Jehle, Wolfgang Küttler (Eds.): Historisch-kritisches Wörterbuch des Marxismus. Eine Veröffentlichung des Berliner Instituts für Kritische Theorie (InkriT). Hamburg: Argument, pp. 1387–1406.

Schrage, Elliot J.; Ewing, Anthony P. (2005): The Cocoa Industry and Child Labour. In *JCC* 18.

Schroth, Götz; Läderach, Peter; Martinez-Valle, Armando Isaac; Bunn, Christian; Jassogne, Laurence (2016): Vulnerability to climate change of cocoa in West Africa: Patterns, opportunities and limits to adaptation. In *The Science of the total environment* 556, pp. 231–241. DOI: https://doi.org/10.1016/j.scitotenv.2016.03.024.

Schutter, Olivier de (2014): Corporate Social Responsibility European Style.

Schwartz, Corina Teodora (2013): RurbanAfrica Policy Brief 1. European Commission (RurbanAfrica Policy Brief, 1).

Schwindenhammer, Sandra (2017): Global organic agriculture policy-making through standards as an organizational field: when institutional dynamics meet entrepreneurs. In *Journal of European Public Policy* 24 (11), pp. 1678–1697. DOI: https://doi.org/10.1080/135 01763.2017.1334086.

Shamir, Ronen (2004): The De-Radicalization of Corporate Social Responsibility. In *Critical Sociology* 30 (3), pp. 669–689. DOI: https://doi.org/10.1163/1569163042119831.

Shamir, Ronen (2008): Corporate Social Responsibility: Towards a New Market-Embedded Morality? In *Theoretical Inquiries in Law* 9 (2).

Shamir, Ronen (2010): Capitalism, Governance, and Authority: The Case of Corporate Social Responsibility. In *Annu. Rev. Law. Soc. Sci.* 6 (1), pp. 531–553. DOI: https://doi. org/10.1146/annurev-lawsocsci-102209-153000.

Shamir, Ronen (2011): Socially Responsible Private Regulation: World-Culture or World-Capitalism? In *Law & Society Review* 45 (2), pp. 313–336. DOI: https://doi.org/10.1111/ j.1540-5893.2011.00439.x.

Shapiro, H-Y; Rosenquist, E. M. (2004): Public/private partnerships in agroforestry: the example of working together to improve cocoa sustainability. In *Agroforest Syst* 61–62 (1–3), pp. 453–462. DOI: https://doi.org/10.1023/B:AGFO.0000029025.08901.9c.

Simon, Roger (Ed.) (1991): Gramsci's political thought. An introduction. Completely rev. and reset. London: Lawrence & Wishart.

Singer, Peter (2015): Praktische Ethik. 3., rev. u. erw. Aufl., [Nachdr.]. Stuttgart: Reclam (Reclams Universal-Bibliothek, 18919).

Sklair, Leslie (1997): Social movements for global capitalism: the transnational capitalist class in action. In *Review of International Political Economy* 4 (3), pp. 514–538. DOI: https://doi.org/10.1080/096922997347733.

Sklair, Leslie (2003): The transnational capitalist class. Reprinted. Oxford: Blackwell.

Slager, Rieneke; Gond, Jean-Pascal; Moon, Jeremy (2012): Standardization as Institutional Work: The Regulatory Power of a Responsible Investment Standard. In *Organization Studies* 33 (5–6), pp. 763–790. DOI: https://doi.org/10.1177/0170840612443628.

Snapir, B.; Simms, D. M.; Waine, T. W. (2017): Mapping the expansion of galamsey gold mines in the cocoa growing area of Ghana using optical remote sensing. In *International Journal of Applied Earth Observation and Geoinformation* 58, pp. 225–233. DOI: https:// doi.org/10.1016/j.jag.2017.02.009.

Social Accountability Accreditation Services (2018): SA8000 Certification. Available online at http://www.saasaccreditation.org/sa8000-certification, checked on 9/26/2020.

Statista (2017): Statista dossier on the chocolate industry. Edited by Statista Ltd. Available online at https://www.statista.com/study/10607/chocolate-statista-dossier/, checked on 4/ 26/2017.

Statista (2020a): Cocoa production by country 2019/2020 | Statista. Available online at https://www.statista.com/statistics/263855/cocoa-bean-production-worldwide-by-reg ion/, checked on 9/1/2020.

Statista (2020b): Consumption of chocolate worldwide, 2012/13–2018/19. Available online at https://www.statista.com/statistics/238849/global-chocolate-consumption/, checked on 9/1/2020.

Statista (2020c): Leading countries of cocoa bean processing worldwide, 2019/2020. Available online at https://www.statista.com/statistics/238242/leading-countries-of-glo bal-cocoa-bean-processing/, checked on 9/1/2020.

Stiftung Warentest (2016): 5 Nachhaltigkeitssiegel für Lebensmittel. Available online at https://www.test.de/Nachhaltigkeitssiegel-Koennen-Verbraucher-Fairtrade-Utz-Co-ver trauen-5007466-5009333/?sort=gesamtErgebnis, checked on 10/1/2020.

Sturgeon, T. J. (2002): Modular production networks: a new American model of industrial organization. In *Ind Corp Change* 11 (3), pp. 451–496. DOI: https://doi.org/10.1093/icc/ 11.3.451.

Sturgeon, Timothy (2009a): From Commodity Chains to Value Chains: Interdisciplinary Theory Building in an Age of Globalization. In Jennifer Bair (Ed.): Frontiers of commodity chain research. Stanford, Calif.: Stanford University Press, pp. 110–135.

Sturgeon, Timothy (2009b): From Commodity Chains to Value Chains: Interdisciplinary Theory Building in an Age of Globalization. In Jennifer Bair (Ed.): Frontiers of commodity chain research. Stanford, Calif.: Stanford University Press, pp. 110–135.

Talbot, John M. (2002): Tropical commodity chains, forward integration strategies and international inequality: coffee, cocoa and tea. In *Review of International Political Economy* 9 (4), pp. 701–734. DOI: https://doi.org/10.1080/0969229022000021862.

Talbot, John M. (2010): Grounds for Agreement. The Political Economy of the Coffee Commodity Chain. Lanham: Rowman & Littlefield Publishers. Available online at http:// search.ebscohost.com/login.aspx?direct=true&scope=site&db=nlebk&db=nlabk&AN= 613589.

Tallontire, Anne (2007): CSR and Regulation: Towards a Framework for Understanding Private Standards Initiatives in. In *Third World Quarterly* 28 (4), pp. 775–791. Available online at http://www.jstor.org/stable/20454961, checked on 4/30/2016.

The Sustainable Trade Initiative (n.d.): About IDH. Driving sustainability from niche to norm. Available online at https://www.idhsustainabletrade.com/about-idh/, checked on 10/1/2020.

The World Bank (2017): Governance and the Law. World Development Report. Washington DC. Available online at https://www.worldbank.org/en/news/press-release/2017/01/30/ improving-governance-is-key-to-ensuring-equitable-growth-in-developing-countries, checked on 10/4/2020.

Theuws, Martje; van Huijstee, Mariëtte (2013): Corporate Responsibility Instruments. A Comparison of the OECD Guidelines, ISO 26000 + the UN Global Compact. Available online at https://www.somo.nl/corporate-responsibility-instruments/, checked on 8/ 19/2020.

Ton, Giel; Hagelaar, Geoffrey; Laven, Anna; Vellema, Sietze (2008): Chain governance, sector policies and economic sustainability in cocoa. A comparative analysis of Ghana, Côte d'Ivoire, and Ecuador. Wageningen (Markets, Chains and Sustainable Development, 12). Available online at http://www.boci.wur.nl/UK/Publications/, checked on 9/26/2016.

Traoré, Doussou (2009): Cocoa and Coffee Value Chains in West and Central Africa: Constraints and Options for Revenue-Raising Diversification. Edited by Food and Agriculture Organization of the United Nations (AAACP Paper Series, 3). Available online at http://www.fao.org/fileadmin/templates/est/AAACP/westafrica/FAO_AAACP_Paper_ Series_No_3_1_.pdf, checked on 4/30/2016.

Trienekens, Jacques; Zuurbier, Peter (2008): Quality and safety standards in the food industry, developments and challenges. In *International Journal of Production Economics* 113 (1), pp. 107–122. DOI: https://doi.org/10.1016/j.ijpe.2007.02.050.

UNCTAD (2008): Cocoa Study: Industry Structures and Competition. United Nations Conference for Trade and Development. Available online at http://unctad.org/en/docs/ditcco m20081_en.pdf, checked on 9/26/2016.

UNCTAD (2012): Corporate Social Responsibility in Global Value Chains. Evaluation and monitoring challenges for small and medium sized suppliers in developing countries. United Nations. New York, Geneva. Available online at http://unctad.org/en/Publications Library/diaeed2012d3_en.pdf, checked on 4/30/2016.

UNCTAD secretariat (2016): Agricultural commodity value chains. The effects of market concentration on farmers and producing countries – the case of cocoa. United Nations Conference on Trade and Development (Evolution of the international trading system and its trends from a development perspective, TD/B/63/2). Available online at http://unctad. org/meetings/en/SessionalDocuments/tdb63d2_en.pdf, checked on 4/29/2017.

Ungericht, Bernhard; Hirt, Christian (2010a): CSR as a Political Arena: The Struggle for a European Framework. In *Bus. polit.* 12 (4), pp. 1–22. DOI: https://doi.org/10.2202/1469-3569.1303.

Ungericht, Bernhard; Hirt, Christian (2010b): Politik-sensible CSR-Forschung am Beispiel der Auseinandersetzung um ein europaeisches Rahmenwerk (Political Aspects of CSR Research – Struggling for a European CSR Framework). In *Zeitschrift für Wirtschafts- und Unternehmensethik – Journal for Business, Economics & Ethics* 11 (2), pp. 174–192.

United Nations (1993): International Cocoa Agreement. Edited by United Nations.

United Nations (2011): Report of the Special Representative of the Secretary- General on the issue of human rights and transnational corporations and other business enterprises, John Ruggie. Guiding Principles on Business and Human Rights: Implementing the United Nations "Protect, Respect and Remedy" Framework. A/HRC/14/27. Edited by United Nations. New York. Available online at https://www.ohchr.org/EN/Issues/Transnationa lCorporations/Pages/Reports.aspx, checked on 8/31/2020.

United Nations (2015): Sustainable Development Goals. The Sustainable Development Agenda. Available online at https://www.un.org/sustainabledevelopment/development-agenda/, checked on 9/26/2020.

United Nations Conference on Trade and Development (2020): Topsy-turvy world: net transfer of resources from poor to rich countries (UNCTAD Policy Brief, 78). Available online at https://unctad.org/en/pages/PublicationWebflyer.aspx?publicationid=2744, checked on 10/4/2020.

United Nations Development Programme (2013): Support for Development and Operation of COCOBOD's Ghana Cocoa Platform. Available online at https://www.gh.undp. org/content/ghana/en/home/library/environment_energy/publication_2.html, checked on 9/12/2020.

United Nations Forum on Sustainability Standards (2013): Today's landscape of issues & initiatives to achieve public policy objectives. Part 1: issues. Available online at https://unfss. org/home/flagship-publication/, checked on 9/12/2020.

United Nations Forum on Sustainability Standards (2017): Policy Brief: Fostering the Sustainability of Global Value Chains (GVCs). In *UNFSS*, 4/11/2017. Available online at https://unfss.org/2017/04/11/fostering-the-sustainability-of-global-value-chains-gvcs/, checked on 9/1/2020.

United Nations Global Compact (1999): The Ten Principles of the UN Global Compact. Available online at https://www.unglobalcompact.org/what-is-gc/mission/principles.

United Nations Human Rights Office of the High Commissioner (1989): Convention on the Rights of the Child. Available online at https://www.ohchr.org/en/professionalinterest/pages/crc.aspx, checked on 9/7/2020.

United Nations Human Rights Office of the High Commissioner (2011): Guiding Principles on Business and Human Rights. Implementing the United Nations "Protect, Respect and Remedy" Framework. HR/PUB/11/04. United Nations. New York, Geneva.

Utting, Peter (2005): Corporate responsibility and the movement of business. In *Development in Practice* 15 (3–4), pp. 375–388. DOI: https://doi.org/10.1080/09614520500075797.

Utting, Peter (2007): csr and equality. In *Third World Quarterly* 28 (4), pp. 697–712. DOI: https://doi.org/10.1080/01436590701336572.

Utting, Peter; Marques, José Carlos (2013): Corporate social responsibility and regulatory governance. Towards inclusive development? Basingstoke: Palgrave Macmillan (International political economy series). Available online at http://www.esmt.eblib.com/patron/FullRecord.aspx?p=578829.

UTZ – Rainforest Alliance (2020a): Certification. Available online at https://utz.org/what-we-offer/certification/, checked on 10/11/2020.

UTZ – Rainforest Alliance (2020b): Cocoa. Available online at https://utz.org/what-we-offer/certification/products-we-certify/cocoa/?gclid=Cj0KCQjw2PP1BRCiARIsAEqv-pTH-ks3D1qf13slskbbvLfpRuDWJTaIcyqRK0cyFb5n_aClPiS5e5gaAIXIEALw_wcB, checked on 10/1/2020.

UTZ certified (2014): Core Code of Conduct Version 1.0. For individual and multi-site certification. Available online at https://utz.org/?attachment_id=21461, checked on 10/12/2020.

UTZ certified (2015): Code of Conduct Cocoa Module 1.1 – 2015. Available online at https://utz.org/?attachment_id=21464, checked on 9/26/2020.

Vellema, Sietze; Laven, Anna; Ton, Giel; Muilerman, Sander (2016): Policy Reform and Supply Chain Governance. Insights from Ghana, Côte d'Ivoire, and Ecuador. In Mara P. Squicciarini, Johan F. M. Swinnen (Eds.): The economics of chocolate. Oxford: Oxford University Press, pp. 228–246.

Vigneri, Marcella; Holmes, Rebecca (2009): When being more productive still doesn't pay: gender inequality and socio-economic constraints in Ghana's cocoa sector. Paper presented at the FAO-IFAD-ILO Workshop. Edited by Overseas Development Institute. Rome.

Vlaeminck, Pieter; Vandoren, Jana; Vranken, Liesbet (2016): Consumers' Willingness to pay for Fair Trade Chocolate. In Mara P. Squicciarini, Johan F. M. Swinnen (Eds.): The economics of chocolate. Oxford: Oxford University Press.

Voegtlin, Christian; Scherer, Andreas Georg (2017): Responsible Innovation and the Innovation of Responsibility: Governing Sustainable Development in a Globalized World. In *Journal of Business Ethics* (143), pp. 227–243.

Voora, Vivek; Bermúdez, Steffany; Larrea, Cristina (2019): Global Market Report: Cocoa. Edited by The International Institute for Sustainable Development. Winnipeg, Manitoba. Available online at https://www.jstor.org/stable/resrep22025?seq=1#metadata_info_tab_contents, checked on 9/12/2020.

Votaw, Dow (1972): Genius Becomes Rare: A Comment on the Doctrine of Social Responsibility Pt. I. In *California Management Review* 15 (2), pp. 25–31. DOI: https://doi.org/10.2307/41164415.

Waarts, Yuca; Ingram, Verina; Linderhof, Vincent; Puister-Jansen, Linda; van Rijn, Fedes (2015): Impact of UTZ certification on cocoa producers in Ghana, 2011 to 2014. LEI Wageningen UR. Den Haag.

Wessel, Marius; Quist-Wessel, P. FolukeM. (2015): Cocoa production in West Africa, a review and analysis of recent developments. In *NJAS – Wageningen Journal of Life Sciences* 74–75, pp. 1–7. DOI: https://doi.org/10.1016/j.njas.2015.09.001.

Wetzels, Hans (2019): Countries propose a treaty to end corporate impunity. United Nations, Africa Renewal. Available online at https://www.un.org/africarenewal/magazine/april-2019-july-2019/countries-propose-treaty-end-corporate-impunity, checked on 10/7/2020.

Wexler, Alexandra (2016): Chocolate Makers Fight a Melting Supply of Cocoa. In *The Wall Street Journal*, 1/13/2016. Available online at http://www.wsj.com/articles/chocol ate-makers-fight-a-melting-supply-of-cocoa-1452738616.

Whelan, Glen (2012): The Political Perspective of Corporate Social Responsibility: A Critical Research Agenda. In *Bus. Ethics Q.* 22 (4), pp. 709–737. DOI: https://doi.org/10. 5840/beq201222445.

Wilson, Edward Osborne (2002): The future of life. 1st ed. New York: Alfred A. Knopf. Available online at http://worldcatlibraries.org/wcpa/oclc/47240796.

Winter, Jens (2007): Transnationale Arbeitskonflikte. Das Beispiel der hegemonialen Konstellation im NAFTA-Raum. Zugl.: Bremen, Univ., Diss., 2006. 1. Aufl. Münster: Verl. Westfälisches Dampfboot.

Winter, Jens (2011): Dimensionen einer hegemonialen Konstellation. Eckpunkte einer akteursorientierten kritisch-hegemonietheoretischen Forschungsperspektive. In Benjamin Opratko, Oliver Prausmüller (Eds.): Gramsci global. Neogramscianische Perspektiven in der internationalen politischen Ökonomie. Dt. Orig.-Ausg. Hamburg: Argument-Verl. (Argument Sonderband, N.F., 310), pp. 145–162.

World Bank (1983): Ghana: The Cocoa Sector. Background Paper 1 of 4 Prepared for the Ghana Policies and Programme for Adjustment Report 4702-GH. Working Paper of Western Africa Programmes Department I, Division B. Washington, DC.

World Bank (2013): Supply Chain Risk Assessment: Cocoa in Ghana. World Bank. Washington DC. Available online at http://documents.worldbank.org/curated/en/151931468151 162220/Ghana-Cocoa-supply-chain-risk-assessment, checked on 11/4/2017.

World Cocoa Foundation (n.d.a): Focus Areas. Available online at https://www.worldcoco afoundation.org/focus-areas/, checked on 10/4/2020.

World Cocoa Foundation (n.d.b): Initiatives. Available online at https://www.worldcocoafo undation.org/initiatives/, checked on 10/1/2020.

World Cocoa Foundation (n.d.c): Our Members. Available online at https://www.worldcoco afoundation.org/about-wcf/members/, checked on 9/26/2020.

World Cocoa Foundation (n.d.d): Vision and Mission. Available online at https://www.wor ldcocoafoundation.org/about-wcf/vision-mission/, checked on 10/1/2020.

World Cocoa Foundation (2012): Ghana Cocoa Board and World Cocoa Foundation Sign Agreement to Continue Successful CocoaLink Program. Accra (Press Release). Available online at https://www.worldcocoafoundation.org/press-release/press-release-ghana-cocoa-board-and-world-cocoa-foundation-sign-agreement-to-continue-successful-coc oalink-program/, checked on 10/12/2020.

World Cocoa Foundation (2014): Cocoa Market Update. Available online at http://www.wor ldcocoafoundation.org/wp-content/uploads/Cocoa-Market-Update-as-of-4-1-2014.pdf, checked on 10/4/2017.

Worth, Owen (2008): The Poverty and Potential of Gramscian Thought in International Relations. In *Int Polit* 45 (6), pp. 633–649. DOI: https://doi.org/10.1057/ip.2008.31.

Wullweber, Joscha (2014): Leere Signifikanten, hegemoniale Projekte und internationale Innovations- und Nanotechnologiepolitik. In Eva Herschinger, Judith Renner (Eds.): Diskursforschung in den Internationalen Beziehungen. Diskursforschung in den Internationalen Beziehungen. 1. Auflage. Baden-Baden: Nomos Verlagsgesellschaft mbH & Co. KG (Innovative Forschung – Theorien, Methoden, Konzepte, 1), pp. 270–306.

Yamoah, Fred A.; Kaba, James S.; Amankwah-Amoah, Joseph; Acquaye, Adolf (2020): Stakeholder Collaboration in Climate-Smart Agricultural Production Innovations: Insights from the Cocoa Industry in Ghana. In *Environmental management*, pp. 1–14. DOI: https://doi.org/10.1007/s00267-020-01327-z.

Printed in the United States
by Baker & Taylor Publisher Services

Printed in the United States
by Baker & Taylor Publisher Services